AF492398

Author's website:

http://www.guillemant.net/english

Original title in French: *La Route du Temps*, Éd. Le Temps Présent.

Talma Studios
231, rue Saint-Honoré
75001 Paris – France
www.talmastudios.com
info@talmastudios.com

Cover picture: 51398335 © De Firefly Designers | Shutterstock.com

ISBN: 979-10-96132-70-6   EAN: 9791096132706

Philippe Guillemant

THE ROAD OF TIME

THE BOOK THAT CHANGES

EVERYTHING WE KNOW ABOUT TIME

A translation from French by Elinor Dorday

*To real intelligent powers
who know that Love is the key.*

To my children Antoine and Lambert.

FOREWORD

When I finalized the first edition of this book in 2009, I would never have imagined that in the years to come, the bold ideas I put forward would be independently taken up by the international scientific community in several publications, books and lectures by mainstream physicists with undisputed credibility. The ideas in question were clearly in the air and therefore *The Road of Time* had made its debut at the right time, when, without being marginalized, concepts such as "The future is already there," "The future influences the present" or "The present influences the past" could finally be taken seriously. These are not yet well-established scientific truths, but these concepts have become admissible because the usual view we have of time today is clearly contradicted by science: reality is not created in the present, for nowhere is there a "front of the present" before which everything would be created and after which nothing at all would happen. This "front of the present" is now clearly considered as an illusion purely linked to our consciousness, and it follows the need to totally revise this simplistic vision of the creation of our reality. This is what I have done and which I reaffirm in this English edition by going even further, in an independent fifth part where I describe a "territory of thought" much wider and steeper than we think, which harmonizes very well with this already very mountainous "Road of Time."

The enthusiastic reception that the first edition received led me to give about fifty lectures since 2010, to write two other books and most of all to introduce the physics of information as a research subject in my laboratory, today recognized by some scientific publications, the last one in the prestigious journal *Annals of* Physics[1].

The new edition of this book remains very current thanks to the updates made in the English version and among these, we will find in the last part the transcript of the lecture I delivered at the Institut de France in Paris, as part of a symposium about consciousness that was organized in 2012 by an association of french academicians. I presented there scientific arguments in favor of the model of consciousness of this book, in particular the reasons why the two physical functions of intention and observation are

---

1. Philippe Guillemant et al, *A Discrete Classical Space Time Could Require 6 Extra Dimensions*, *Annals of Physics 388* (2018), 428-442.

really shaping space-time, that is presiding its atemporal evolution in the future and the past.

The first two parts aim to analyse delicate concepts such as the irreversibility of time, an irreversible observation, in my opinion, in order to lead the reader to radically transform his vision of time through the most implacable possible reasoning, so that he can find rational ideas such as "traces of the future" and especially the idea that our thoughts, in particular our intentions, necessarily have an influence on the creation of our reality, long before our actions. Above all, I wanted to show that it is not necessary to call upon quantum mechanics, still too obscure for the average person, to understand that we do indeed create our reality through our thoughts, but that this construction concerns our future and not our present, contrary to what a naïve interpretation of the role of the quantum mechanics observer, exploited by the "New Age" thought, would tend to make us believe.

The third part of this book is devoted to the description and critical analysis of my personal experiences of strange coincidences and especially synchronicities, the latter resulting from a real experiment that I decided to carry out on myself after having discovered why and how it was possible to provoke them. I have drawn from it a kind of practical and spiritual guide, modelized in the fifth part through the physical functions of consciousness and intended for the reader who will want to try to reproduce my experiment, which has already been successfully done by many of the first edition readers. The fourth part finally draws the consequences of the theory of double causality on our vision of space-time. The result is a true revelation about our creative role in the universe, a role that implies that our primary nature is spiritual and that this essence, love in the purest sense of the word, is not a product of brain chemistry but something even more fundamental than gravitation or light, external to our space-time and directly related to our free will.

Paradoxically, because of the upheaval of its vision of time, physics is today confronted with the inevitable esotericism of a materialistic vision of the world which claims that man's free will is illusory and that consequently, our consciousness could in no way directly influence the course of future events. This is a purely mechanistic belief which leads the physicists who cling to it to affirm even more unthinkable things such as: there would exist myriads of parallel universes where we could have an astronomical quantity of duplicates of ourselves completely conscious; or else: our universe would be a space-time forever frozen which was created *all at once*, from the Big

Bang until the end of time, that we only visit without being able to change anything there. Today the coherent materialists are thus stuck between, on the one hand, the extravagance of their vision of the multiverse and, on the other hand, the creationism of their vision of space-time. It is therefore high time to drop the dogma of temporal determinism, a kind of scientific machismo that has become irrational, but which has nevertheless locked us little by little into an absurd "park of thought", which could one day transform us into machines if we are not careful, even into human cattle.

The current crisis of our society is thus, undoubtedly, in my opinion, directly linked to the sick end of a materialism that exhibits its last rales, like that of our illustrious Stephen Hawking (1)[2] when he claimed that philosophy would be dead, that we would be machines and that we would not need God. The good news is that these ideas are no longer accepted and that we can count, I am sure, on the scientific community—at least the French-speaking one—to reverse this outdated vision and give birth to a new paradigm that will profoundly transform man and make him discover his spiritual nature. Of course, this will not happen in one day, but more good news is that we may not need to wait for new discoveries in physics to accelerate this liberating movement. Already, the new physics of information that has just been brought to light brings us to the threshold of maturity of new computer technologies that will force man to realize through experience the creative role of his *consciousness*.

As Jacques Vallée, who convincingly bases these new physics on the double causality[3], testifies, one of the primers that have already begun to catch fire will undoubtedly contribute to the explosion of this new paradigm in the coming years. This beginning has its origin in this book: it is about the development of the intrinsic capacity that we have to provoke strange coincidences, extraordinary accidents that introduce magic into our lives and that could go so far as to re-enchant the world when we finally assimilate their nature. Thus, this book is above all a hiking trail to guide true rationalists outside the current park of thinking, to lead them to contemplate the true territory of reason, then finally discover spirituality by experiencing synchronicity.

---

2. For figures in brackets, see Bibliography on page 276.
3. Jacques Vallée, *A Theory of Everything (Else)*, A lecture at TedX Brussels, November 22, 2011.

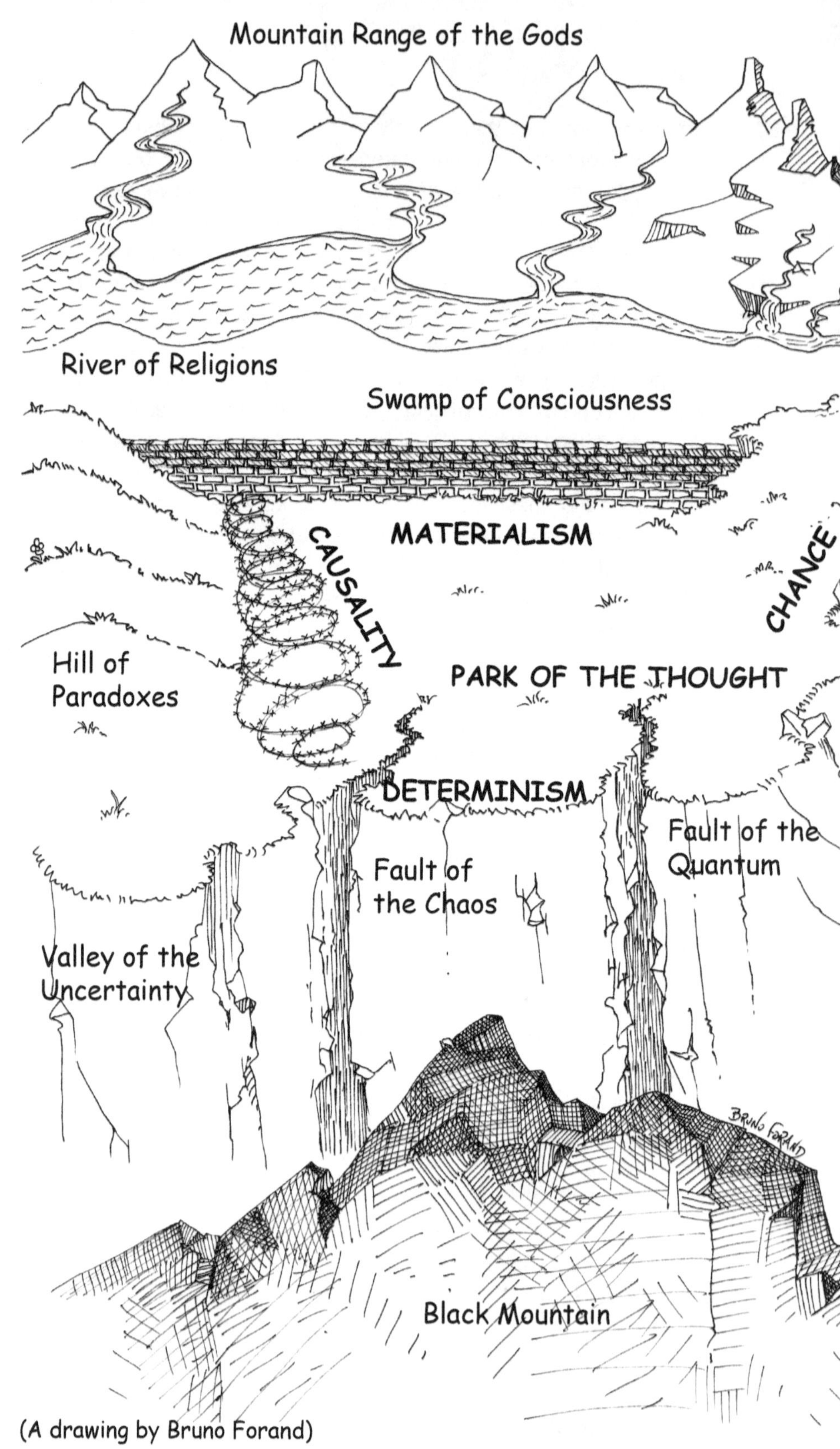

(A drawing by Bruno Forand)

Peak of the Spirit
Road of Time
Pass of the Angel
Path of the Free Will
Gorges of the Creation
CHANCE
Ford of the Purpose
Ravine of the Creation
Waterfall of the Intention
Landfill of the Chance
Abyss of the Illusion
The Territory of the Thought

INTRODUCTION

*In which we suggest that instinctive dialogue with nature
can help us exercise our own free will.*

The Road of Time is a little mountain road that winds endlessly through the Geological Reserve of Haute-Provence, from the medieval city of Sisteron down to the Asse Valley, passing close by the city of Digne. Apart from its intrinsic geological value and the breathtaking beauty of the surrounding landscape, there is something particularly interesting about this road, because historians believe that it leads to the city of Theopolis, otherwise known as the "City of God", the famous city of St Augustin that remains undiscovered to this day, in spite of numerous archaeological explorations carried out in the area: their conjecture is based upon information contained in a 5th century Roman inscription carved into one of the rocks on the "Pierre Ecrite" mountain pass, which forms a sort of "entrance hall" into the Geological Reserve.

The mystery surrounding the Road of Time is only one of many attractive features of the enchanting landscape: stunning fossils, mysterious rock formations, "magic" stones, hidden yurts high up in the hills, astonishing wildlife and enigmatic natives with personalities thrown into sharp relief by the area's outstanding natural light—each with its own peculiarity that lends a sense of mystery to the path. It is all the more marked because the rare encounters we make seem to hold some sort of meaning, as if we were in a role-playing game where each actor had clues to offer us for our onward journey.

This low mountain range, where altitudes vary from five hundred to two thousand metres, is exceptional because of the huge wealth of possible routes it offers through the reserve, confronting us with a multiplicity of paths to choose from. Hiking here, I have felt how one's inner strength is restored by these kinds of walks, where you never quite know where you are going, or what you are going to find.

Generally, by taking the steepest way up, and after several hours of climbing, during which my mind would finally manage to free itself from any physical tension, I would at last find myself becoming receptive to the nature around me. And then nature would actually take over from my body and answer my inner questions, like a sort of echo of my mind. This new-found sensitivity would prove ample reward for all the effort it had taken to get up there, and each time the feeling would come over me quite unexpectedly. We often forget the virtues of hill-walking, which go far beyond mere physical exercise—it helps the mind to refocus in a sane and liberated environment. Both inside and out, altitude lifts us and broadens our vision: the meaning of our life becomes clearer or may even reappear, as if nourishing our intuition this way allows us to let it take over our mental reins once more. Nature offers us a way of learning how to circumvent difficulties, both inner and outer ones, by revealing an entirely unsuspected range of possibilities to the "living beings" we are; like a challenge we have to seek within ourselves before testing it upon the training ground of intuition, as we search for the right attitude to adopt on our progress through real life.

Even more so than altitude and visibility, it is the characteristic solitude and difficulty of hill-walking that makes us refocus on ourselves, lending that sort of nobility to our inner life which comes from journeys of initiation. Of all difficulties, the uphill climb is the one which best compensates for the effort it demands of us. Other obstacles symbolize age-old problems, from the most commonplace to those that take the longest to overcome: and yet here in nature, the rocks, thorns, dense forest, waterfalls, rocky outcrops and ambiguous paths all offer quite unsuspected possibilities for overcoming them—as long as we look out for them. When we do, they act as so many invitations to find new ways of adapting in real life. This low-lying, mountainous country teaches us an unexpected lesson, probably because of its richness and diversity: nature shows us that every problem contains its own solution, providing we are ready to admit that the problem exists and listen to what it means, so that we can then contemplate with serenity the other possible paths open to us.

Is this, then, the true virtue of the mountain range, to be constantly offering us the gentleness of multiple compromise from within its large spectrum of nuances? Or is it the majesty of the place and its profound silence which make us feel that a third person is watching our progress, always ready to suggest some new direction? In the worst or best of situations, an unexpected encounter sometimes invites itself into our reflexions, seeming to act as a signal that one phase is ending and another must be overcome. During this encounter, our plans may suddenly alter, or foresee some future evolution, or merely absorb information which will be useful at the next fork in the path. And then, almost like confirmation that the encounter was planned in advance by the universe, the choice to be made at the next fork in the path will appear obvious, as if someone had changed the points at the junction. Sometimes, it may simply be the weather which either guides or restricts us through encouragement or dissuasion, simulating the conditions for our decision according to our mood. Or else "signs" are sent directly by nature, optical illusions formed by shadow and light or dynamic effects, noise made by the wind or passing animals, appearing to us like warnings, invariably influencing our journey, although we would never seriously admit as much to ourselves.

Often, in fact, it is the chance meetings that direct us along the Road of Time: encounters with talking stones (17), natural phenomena or living beings. The same chance that is responsible for altering our real lives, when we meet our future partner, employer or friend. But why is it that we have to wait until we are wandering through the countryside, before we can discern the hidden meaning of the most significant encounters of our lives, the meetings that lend new meaning to chance? Could it be that it is easier to comprehend things when we are involved in a lighter, more detached sort of way?

Having experienced all these encounters and adventures on the Road of Time as so many symbols or signs that helped me gain a better understanding of my whole life, while at the same time feeling I had come through a period of personal evolution, I gradually came to realize unexpected potential in terms of my own free will. Was it the transposition of ordinary existence, where I seemed to be confronted with major choices almost daily? Whatever the case, I developed an ability to observe elements of my surroundings that had nothing to do with my own problems, but which nonetheless allowed me to reach a state of awareness that helped make them go away. Sometimes it happened quite simply, as soon as their critical aspect disappeared. Better yet, though, whenever these elements appeared to act as an assembly of constituent parts talking with the same voice, conversing in the language of coincidence, I would reap a harvest of suggestions that would help me create my own life map.

Just as nature's marvels and obstacles invited me to listen to them as I walked, holding up a mirror to my inner questions, even forming an extension of my own self, I also discovered that by listening to the events in my real life, rather than merely submitting to them as if they were constraints, new and unsuspected paths would open up before me. Was this an alternative practice to making rational but dubious choices, or was it heightened personal intuition? I couldn't say, except that the experiences I went on to have as a result far surpassed everything I might have hoped for in terms of benefit, and it was then that I realized that I had to share my luck.

The first manifestation of this luck came in the form of a series of extraordinary coincidences that befell me on the Road of Time, apparently acting as confirmation that I had taken the right path. They led me to think long and hard about the meaning of time, something I had been pondering for many decades, bringing me to an intuitive sensation of communing with the elements of the universe, as if my brain were reaching far beyond the limits of my own self. I went on to understand how this sensation, which is a source of happiness and confidence in life, might well come from something other than a mere illusion. And so, little by little, I began to take apart the mechanics of such mysterious things as the phenomenon of synchronicity* (see Glossary on page 263). In this book, I will be exposing these mechanisms using a double-edged scientific and philosophical approach, underpinned by what is intended to be implacable logic. The reader may be surprised to observe that science, poetry and spirituality can be excellent bedfellows.

The Road of Time is such an effective and natural model that I cannot employ it too liberally to explain the fundamental things in life it has helped me to understand, culminating above all in the ability to reconcile science and faith. The Geological Reserve will provide us with metaphorical and symbolic terrain on which to express our mental environment, so that we may explain and exercise our inner potential, thanks to a set of powerful analogies corresponding to mental properties: not just the altitude, remoteness or difficulty of the walk, but also the presence of water, direction, temperature and weather conditions, as we will be seeing later on.

In this book, we will reveal unheard-of mechanisms for exercising the freedom we possess for "programming" our future, a freedom that exploits little-known intellectual properties and which it is infinitely more judicious to understand using this type of metaphor* than by any pre-established psycho-scientific approach.

We will show that we have unexploited potential deep within us, and in particular an instant ability to activate that future which best corresponds to our intentions, from among the multiple possible paths of our life. And yet, imprisoned as we are within our everyday conditioning, we are generally unable to exploit this potential, because our decisions are hampered by all sorts of misconceptions—about our minds and the idea we have of ourselves—and suffer interference from various determinisms, especially habit.

Thus the choice of countless paths we can follow through the Geological Reserve will serve as a metaphor to mimic, comprehend and even experiment on the mechanisms we are about to reveal. First of all, to simplify, I will sum up these possibilities using the model of the "Tree of Life"*. We will see that most of the time in our lives, we do not experiment with the wealth of our potential choices, and even that we seldom venture off the beaten track, so assailed are we by preconceived ideas (clichés, fixations and all sorts of preconditioning): most of the time we keep to the tarmac, the "main trunk" of our life.

In this essay on time, I will be introducing, in entirely rational terms and for the price of a first, steep section about the reversal of time and non-causal logic, the astonishingly little-known mechanism of a truly magical element of life, by suggesting a new Theory of Time, called the "Theory of Double Causality*". This theory will then be put into practice and tested against experience, the Road of Time having been chosen not only as a testing ground for metaphorical simulation, but also for real-life experiments.

I will not be calling on any esoteric references, but simply on pure logic and the latest scientific progress, and particularly modern physics.

For the Theory of Double Causality presents primarily scientific interest, because of the way it casts light on the oddest results of the theory of relativity*, statistical physics*, chaos theory* and quantum mechanics*.

Each of these peculiar branches of physics will shed its own light on the subject of time, based on the principles of irreversibility*, causality* or determinism*, the philosophical reaches of which are essential to allow man to find his place in the universe. Looking at it like this, it must be admitted that these principles have always played a reductionist role, even to the point of denying the existence of free will*.

And yet, **the authenticity of free will** is the most fundamental hypothesis of the Theory of Double Causality, to which we will add a second, and no less fundamental hypothesis, namely that **the universe already exists**.

This immediately raises the following question: how can we remain free to act in a universe where our future already exists?

The answer that upholds the Theory of Double Causality is that of the "Tree of Life*": our future already exists in multiple versions, or branches, all coexisting simultaneously and ubiquitously, but as "potentials*" which have not yet been lived out.

Armed with these two fundamental hypotheses and the undeniable result they lead to, we will demonstrate in this book that time possesses virtues we can in all honesty qualify as "magical".

The reference to "magic*" that I employ when talking about "mechanisms" of time may well cause some surprise, but considering its extraordinary properties, no more accurate term could be chosen to qualify the marvellous potential of our life adventure as the universe invites us to live it, if only we are willing to raise our minds and sensitivity to the level of its own subtle laws, and so consent to become the magicians of our own lives.

Part I

REVERSING TIME

I.

THE TREE OF LIFE

*In which we show how the course of our life can be depicted
by a personal Tree of Life, whose branches represent
all the potential choices available to our free will.*

When I reached the top of Mt Mélan, I sat down on a rock and contemplated the extraordinary view below me, which encompassed at least half of the territory travelled through by the Road of Time, the metaphorical territory of multidimensional space* upon which my Tree of Life, like so many others, was inscribed. From Pierre-Ecrite Pass right down to the Thoard Valley, the sun-drenched mountains and valleys that were finally revealed seemed vast enough to symbolize the potential not just of my own life, but of an infinite number of trees.

I waited for my friend Christian, who was taking the climb at his own pace. During the conversation we had just been having about the Tree of Life, he could not help thinking about it in terms of a tree of "reincarnation". Not that he adheres particularly to the idea of reincarnation, simply that the concept of "potential lives" I had tried to explain to him did not suit him. For him, each reality had to be capable of being lived.

I stretched out on the rock to allow a state of meditative slumber to overtake me. My brain began projecting a film before my eyes.

Imposing themselves onto this clear, majestic vision of the countryside, far away to the south and east, came the two branches or paths of the main road, running from the narrow pass near Sisteron, down to the valley that stretched away towards Digne. Behind me, to the north, I could see the countless meanderings that nature had forced on the road. I imagined two types of scenario for the "time tourists" arriving from the direction of the mountain pass, intending to go for a walk or have a picnic. Since even the most modest summit took hours to climb, most of them would abstain from such a hike, never venturing very far from their cars.

In my first scenario, already weary from the endless twists and turns in the road, they would get no further than the Col de Fontbelle. Nevertheless, they would make the most of the short, youthful adventures that would enrich their lives on the nearby footpaths, before going back to civilization for good.

17

In my second scenario, they would be in a hurry to forge straight ahead, determined to succeed in life, making the effort to get beyond the Col and push on as far as the Forest Refuge to obtain their highest academic honours, and then they would descend laboriously back towards Digne, not taking the time to turn off here or there to explore any of the footpaths. They would not leave their car until they got back home, careful to empty the boot of any arms or luggage they had collected like so many acquired habits, preconceived ideas and all sorts of guiding principles they would never manage to be rid of.

I visualized these two possibilities as the two main scenarios for the Road of Time, scenarios in which, starting out with even chances, two completely different types of life would be mapped out.

Yet whatever the chosen scenario might be, I realized that in order truly to explore this territory, one would have to accept the night-time challenge.

In my best-case scenario, I imagined a selected test, a night out camping for example, or in a bed and breakfast or guest hostel, planned in advance for maximum comfort, a choice dictated by reason; or otherwise the bolder choice of bivouacking for the night, in a yurt or a make-shift camp, leaving chance encounters and the lie of the land to dictate events, which might increase people's chances of understanding the mysteries of life and receiving the keys to genuine happiness.

Lastly, in my worst-case scenario, I envisaged the ensuing ordeal if, having failed to anticipate the adventurous nature of a perfectly serious quest, some incident or breakdown should eventually do them the favour—all the more painful for their being unaware of it—of being forced to spend one or more nights there, so that they might realize what they had decided to do with their lives before they were born.

As I tried to understand whether the intention to exist before one had even started to live could make any sense at all, I finally wondered whether this final scenario was not in fact the best of all!

I imagined how, starting out with birth certificates issued from Pierre-Ecrite Pass, the ultimate aim of all these lives must, without doubt, be founded on an identical intention to set out to find the City of God, or, failing that, to live in hope of finding it, knowing that this hope was the foundation of the Tree of Life and the Road of Time.

And I told myself that this hope alone, along with all the original adventures that it would initiate, might even eventually make the ultimate aim of this quest almost secondary.

This is how, failing to remember the true meaning of our own life, I irremediably came to see the Pope himself enthroned with all his retinue, for the purpose of reminding us that we are children of God.

It followed that, having been born at Pierre-Ecrite, most visitors on the Road of Time would finally shake off their quest all the more readily because, as soon as they started their first studies at Saint Geniez, they would be told where to find the sacred place of pilgrimage that would be sufficient to make good men of them.

Here in this unavoidable village, where lots of contradictory information habitually circulated about everything and nothing, particularly as to the supposed site of the City of God, it was difficult not to be told about the most influential sayings and writings, which gave its location as being somewhere near the chapel of Dromon.

All it needed was one apparently serious book on the subject to be published, indicating the location, and fervent believers started making pilgrimages, like converts to a religion

claiming to know God, dreaming perhaps of seeing Theopolis appear on the designated spot.

For the Road of Time conceals a true mystery, one far greater than the small chapel barely large enough to hold a single crib, looking out only onto nature in its wildest state: there is no city anywhere to be seen!

Could the city simply have vanished from the spot as if by magic, like some tiny Atlantis, destroyed and buried beneath the strange heathland to be found at Dromon?

Swept along on this myth, the simplest and most frequently lived life for a time tourist arriving in Pierre-Ecrite Pass consisted in following the main road between Sisteron and Digne, and coming to a halt just opposite Dromon. There, with a single sweep of the eye or a single click of his camera, he would rejoice in having achieved his main objective, before proceeding back down by the most direct route and living out the rest of his life at the lowest altitude.

Looking at the heathland with its extraordinary contours, the imagination can easily satisfy any faith and create a feeling of absolute plenitude. Stopping here has the advantage of relieving one from the adventurous experiences involved in exploring other paths, and which would entail a solitary quest for the road leading to one's own idea of (the city of) God. Were such fear of living or such ignorance about the meaning of life to become generalized, it could result in the same life being systematically relived over and over again, always according to the same scenario, with insignificant variations, such as strolling off to the left or right to sniff the air and pick a few mushrooms.

At this rate, the Road of Time's Tree of Life could swiftly be reduced to a poor imitation of a skeleton, with only parts of the spinal column remaining. And one could hardly recommend that a new visitor with any more highly evolved potential should be born into this Tree of Life, unless he be willing to risk spending his life clearing the way for his successors.

But the quest for the straight path, the fear of living, or merely sightseeing are far from being the only ways of making skeletons of our trees of life. Laziness, social conditioning and the quest for security all have the same tendency to strip our lives bare, without, in the meantime, actually ridding them of any pitfalls.

Thus some of us may accidentally end up at the bottom of a ravine a few hundred metres below the road, like a manager who loses his job and ends up on the dole with no entitlement left and not believing in anything anymore, not even in the "luck" of finding himself down and out. Other victims of life whose faith, on the contrary, will have been pushed to extremes, might find themselves in one of the numerous surrounding monasteries, for the vast, unspoiled open spaces of this mountainous region are highly propitious to the founding of spiritual retreats. Others, better educated without actually believing in the existence of Theopolis, will head straight for maximum safety, their fear mounting with every bend in the road, not really believing in Theopolis, and will end up in Digne in one of our civilization's (retirement) homes, feeling reassured.

Some of the most ambitious followers of the straight path will manage not to get caught up in civilization and will retire further south, to the foothills of the mountains in the Gorges du Verdon, believing they have done their utmost to move closer to Theopolis by arriving at one of the outreaches of the main road. For we do not know exactly where the road ends. All we know is that it carries on somewhere towards the south of the Alpes de Haute-Provence, and that several villages claim to form its furthest point.

And yet, in order for our lives to be successful, nothing obliges us to follow this road to its southernmost point, except society. Society makes us lean towards such a simplistic plan. Many other routes cover even greater distances using a much more complex itinerary, but our social conventions ignore them, favouring instead one royal highway, the one which allows us to travel swift and far.

Finally, whatever the scenario that makes us flee, forget the vertical dimension* and head for the plains, our life will be characterized by the absence of authentic freedom and any sense of "luck". We will become avid for amusement to compensate for our resulting boredom, and we will continue to seek this amusement elsewhere than in the quest for Theopolis. And so we will see that all ways of life that save us from having to face up to the unknown, to life off the beaten track, only serve to justify the myth of Theopolis the unattainable, and to replace it by a creation of our own imagination.

A mouflon, which had been inside my field of vision but obscured by my thoughts, suddenly moved, tearing me away from my imaginary world and bringing me back to reality. Luckily I was quite still, hidden by a few leafy bushes, and the wind was in my favour. The mouflon could not see me. Gently, I took my camera out of my rucksack and started to focus. All at once, even though he had been grazing happily up until then, he suddenly stood up tall and straight as if sensing some threat of danger.

Just as I was about to take the photograph, he bounded back and moved out of range. Missed! I was quite convinced that the mouflon had mistaken my camera for a rifle, and guessed my intentions.

When he had completely disappeared, I saw Christian arrive, winded by the climb. With the perfect bad faith of someone consciously lying to himself, I said:

"You made me miss a mouflon. He sensed you coming and ran away!"

"Oh blast! But you know what, look over there! I just saw a flock of chamois."

He pointed towards the top of Cluchette, but the flock had disappeared. As he came nearer my rock, apparently intending to walk round it, he seemed fascinated by the swelling sea of clouds that had collected at the foot of the valley, along the Vançon River.

"Oh! How beautiful! Tell me, Philippe, your Tree of Life story... Don't you see, it's a dead end here! We can't go on! Our life is finished!"

I was still scrutinizing Cluchette, despite the sun reflecting mirror-like off the rocks on its overhanging cliffs, sometimes to dazzling effect. Christian kept coming nearer, apparently unaware that two metres further on the ground fell away in front of him, plunging into another void. It would have been perilous for him to come a single metre further on his way round my rock.

"Stop! No further, or you'll kill yourself!"

He drew back sharply, then sat down on another, safer rock.

"Goodness! That's some dangerous rock you're sitting on! What did I tell you—the only thing we can do now is commit suicide!"

I was going to have to defend my model of time.

"Of course not, because we're going to go back down."

"Along the same route? That would mean going back in time!"

I was expecting this response, and was happy that our conversation had caught Christian's interest. For me, he represented an audience capable of answering back, for there was a very rational and pragmatic side to him as well as a real sense of spiritual openness, right at the opposite end of the spectrum. In fact, he gave lectures on Buddhism.

I replied, "Not at all! Even if we go back the way we came, life goes on in the same direction, but in a completely different way. It's just that you don't renew your experiences, that's all."

As I went on explaining, Christian's face lit up. Now in retirement, the great summits of his own life were behind him. I explained to him that according to this Theory of Time, his life would be much easier now, proceeding without effort: his journey down not would not force him to live through the same things as before, he would retain the strength of his experience, his sense of having done his duty, and would revisit the same levels but with a different role to play, even if it were only that of informing those he met coming up on his way down. I then reminded him that there were many other ways of re-descending and preparing for retirement, and that the important thing was to avoid descending too low and disappearing into the uniform mass.

"You mean that all your parallel lives since you were born at Pierre-Ecrite have to be lived with variations each time?"

"No, no, not at all, I never said anything about reincarnation!"

Christian had trouble understanding my tale of potential parallel lives, and would rather I had spoken of reincarnation. I had explained that the theory of my book was founded on the hypothesis that our physical, four-dimensional space-time continuum* was contained within a much vaster, multidimensional universe, where we could have numerous parallel lives. This hypothesis was also one of the conclusions of String Theory*, the most elegant—but untested—theory physics has yet produced. The hypothesis is also, according to many physicists, the best interpretation we can make of quantum mechanics, independently of string theory.

To make it easier to represent such space in a way that most closely represents its effect on our human life, the metaphor of the Road of Time had imprinted itself on my mind as being a richer alternative than the diagram of the Tree of Life. But I needed this schematic simplification, representing our life by means of a tree of possibilities, in order to define the potential of life concerning each of us personally.

Seeing Christian's doubtful look, I made a concession.

"Fine, but if you believe in reincarnation, you could always imagine to yourself that other people can be reincarnated in your own life! That way, you would be nearer the truth."

As I was saying this, I was thinking to myself that assuming I really did become reincarnated, then I certainly did not want to have to spend my time building my own tree of life! According to me, it built itself! And it did not even need any help from anyone else, since the various potentials of its future composition were automatically created in the present according to the choices spread out before me.

However, I could not entirely ignore the hypothesis that my present might effectively be "played out" again, for my free will gave substance to this hypothesis, by making each "incarnated" existence into a new opportunity to modify my tree of life, the way a real tree grows.

The theory was not necessary for my hypothesis, though: I could very well have just one life, my own, lived just the once, but my Tree of Life could still nevertheless develop fully by reflecting, moment by moment, all my multiple potential existences already "memorized" in space!

Nonetheless, it is understandable that I should find it difficult to communicate this notion of potential lives without falling back on the hypothesis of reincarnation. Not that it should be definitively cast aside, simply that this theory has no actual need of it. Although it may be interesting, the idea of being able to "replay" a life which already exists gets in the way of understanding the existence of multiple potentials which have not been lived but are still absolutely real.

Christian, however, had no desire to hear about these potentials. By giving him an answer that tried to be broader, I had captured his attention.

"Now you're starting to interest me! Now I'm with you!"

My concession had borne fruit, and so I made it more precise.

"Try to forget about reincarnation. I'm not against it, but I just want to explain how our Tree of Life exists, even if no life has yet lived on it, and even if no one has yet been incarnated on it."

But Christian had stopped listening. Concentrating on the concession I had just made in favour of reincarnation, he was busy simulating his own life choices after leaving the mountain pass.

Attempting to anticipate his thoughts, I began telling myself that each branch of the tree, each fork on the Road of Time, could perfectly symbolize potential choices, no matter whether or not one knew whose life it was, yours or mine, or whether it had ever yet been lived or on the contrary, had already been lived numerous times with an infinite number of variations. The only thing the lives had in common was their birthplace, always here in the mountain pass, for Pierre-Ecrite was the only compulsory part of the route, the main trunk of the Tree. It needed only a single "reference" to be lived at this point for all the others to be brought into existence.

Here in the place I called the Gate of Time, our own birth was greeted by the imposing contours of the land, a change in the surrounding vegetation and a sign telling us that we were entering the Geological Reserve of Haute-Provence. Then we continued climbing, through a valley which led us to the first hamlet along the road.

Christian pointed out to me that a series of amusing coincidences made it seem as if everything in the hamlet was organized to attract young children: an old stable which looked like a nursery, a park with animals...

"It's funny, your idea. It works! And then studying at Saint-Geniez! Ha! That explains why there's so much fuss about the school!"

What interested him, however, was the first fork in the road after the village, inviting him to make a choice.

"So, if I carry straight on and keep going, my life becomes the one I chose, the life of an aviation pilot. But if, when I come back, I turn off to the right towards Sorines, then it's not actually me anymore, because if it had been me, I wouldn't have made a different choice! Because otherwise I wouldn't have gone on to do my studies! Is that what you mean?"

"You could put like that!"

Christian's comments were starting to reach a certain level of credibility. Here, he was asking an awkward question. For any eventual "growth on our Tree of Life" seemed inherent to support reincarnation.

Could it be inevitable that our Theory of Time should lead us to such a conclusion?

The multiple lives susceptible of being "lived" on a Tree of Life seemed to lead to this conclusion by asking the following questions: could it be possible that all the lives on our tree had already been lived, rather than appearing as potential lives? And if that were indeed the case, by whom had they been lived, and when would they be lived again?

Obviously we cannot live all the lives on such a tree simultaneously, and only one of the lives corresponds to our destiny. However, it would be trickier to claim that we are reincarnated on this tree or that someone else would be. This is not what we are proposing, and in this book we will try to restrict ourselves to a "minimalist" version of our Theory of Time, by eliminating any unusable hypotheses, and above all any esoteric idea that might tend to impose itself too swiftly as a shortcut to a system of beliefs, something incompatible with any true spiritual ascension.

So, while not actually excluding the idea of reincarnation, it is in any case unnecessary to our Theory of Time. Moreover, scientists opposed to this idea can easily accept the hypothesis of the Tree of Life based on the theory of parallel universes* (7).

This is why our minimalist model, which makes us set aside notions of "incarnation", does not even oblige us to state that our life could only be lived once, without ever being modified. Nothing in modern physics rules out the possibility that my life or yours may already have been lived a considerable number of times, with or without incarnation! But to prevent the existence of all these lives from becoming incomprehensible or confused, here we will consider that, among all the already existing potentials, only one of our lives is updated* in the present, the one which corresponds to our destiny. For all moments in time coexist simultaneously as constituents of the "reality" of the universe which we conceive as being unique. Whereas the immanent existence of a single universe outside time supposedly leads us to the existence of a destiny which has already been mapped out for each of us!

In order to maintain our free will, therefore, our Theory of Time will be founded on the hypothesis that we are at liberty to modify this destiny at any moment by choosing our path from all the alternative potentials which accompany it. In this way, we would be making our "updated" universe evolve throughout eternity, life after life towards another universe, while we ourselves would be changing our life moment by moment. Eternity and time would be two quite distinct things. Only eternity would be true time, while our own present would be merely a "window" in which all the alternative potentials of our tree of life were updated.

The course of time would thus allow our "real" universe to update itself from among all the potential universes for which it would be the reference point. Thus it would contain all the already-created existences of our trees of life which could be offered to our free will and to our life choices.

But do we really exert such free will and such responsibility within the universe as a result? Is it not, after all, infinitely simpler and space-efficient to think that this supposed freedom of choice is merely an illusion, and that the universe does not care in the least about our presence here?

Chapter 1—Summary

Our theory of Time is grounded on two hypotheses:
- our free will is authentic,
- our future already exists, but may be modified.

The omnipresence* of the future is translated by the existence in the present of multiple potentials of realization, symbolized by our Tree of Life, and hosted by multiple universes.

Among these potentials, one is favoured because it corresponds to the real universe and is where our most probable destiny is situated.

The question as to whether this destiny has already been lived or not is not one that will be addressed in this book, which takes a minimalist approach.

II.

INVERTED DETERMINISM

*In which we see that to maintain our free will,
we must fight against the dogma of irreversibility!*

"You see that pass there? That's where we're going."

"Glad to hear you say so! See? You can read the future!"

"What are you talking about? Oh, yes... but no, this is different, we can see it, it's in the present!"

"Not at all; what's in the present is simply a glimpse of the pass, a trace. When we actually get there, then we'll see things completely differently. We can just about imagine them ahead of time, thanks to that trace."

All of a sudden, clouds and mist began to descend from the Monges Hills, making our goal invisible. The bad weather raging to the north of the Alps was threatening to spread to the Geological Reserve, in spite of the strong Mistral wind busy clearing the skies, breaking up the few clouds which were venturing south.

"Looks like we might get a soaking up there!"

"Hold on... I hope you're not erasing our future!"

I was having fun illustrating what I had been explaining the previous evening. Patrick resisted everything I had told him about time. All I wanted was to make him doubt a little.

"No," he retorted, "we'll keep going, it'll clear. All we have to do is stop and have a drop to drink, seeing as it's nearly lunchtime. Then we'll see how it turns out."

"Wise decision there from your free will!"

"Oh, I can see you coming, you and your blasted free will. If I'd been a robot, I'd have made the same decision."

"Except that you're conscious of your decision, whereas the robot has no need to be conscious to decide."

"So? Whatever we do, it's a brain making the decision, and there's always a reason for the decision, whether it's conscious or not."

"No, we can act for no apparent reason too, using our feelings or emotions, for example."

"But it's still being controlled by your brain. The real reason is still in there somewhere, even if it isn't a conscious one."

"Yes, but not always, otherwise that would mean we'd only have one future, and if that were the case we'd see traces of it."

"There you go with your traces again... but why do you want to see traces of your future? It's nonsense!"

"Because I can see traces of my past, and the laws of nature are reversible in time. The only thing that can explain why we can't see traces of the future is that it is multiple or likely to change at any moment. That's what forces the future to remain indeterminate or invisible, fuzzy or foggy, exactly like that summit we can hardly see any more."

"What are you talking about? It's not true, the laws of nature aren't reversible. Look, there are lots of things that aren't reversible, for example, when you heat up food, or here, if I throw this stone in the river, it's hardly going to come back up again on its own, is it!"

"Yes, I know, they tell us that's because of entropy*, which is constantly increasing. But you know what, that wonderful entropy they taught you about at school is actually entirely empirical, it's just a question of probability: things which are irreversible are only apparently so, and only because it would be infinitely improbable for them to go backwards."

"So? I don't see what difference that makes."

"What difference it makes? But my dear friend, it makes all the difference in the world! Because if everything is reversible, then that means we can just as easily use traces of the future as traces of the past to understand what is happening in the present. And that means that our present itself can actually be a result of our future!"

"But that doesn't make any sense! And what's the point, anyway? The only point for us is to be able to predict the future."

"Listen, we happily use our observations of the present to deduce the past, by trying to understand what has happened. Well, the point of deducing the present from the future is exactly the same thing, to be able better to understand what's happening in the present."

"I don't get it. It's the future that has to be deduced from the present, not the other way round. The present can only be deduced from the past, and anyway, while we're at it, seeing as there's only one past, there can only be one future."

Patrick was busy preparing a home-made cocktail to a secret recipe. Taking a sachet out of his bag and cutting an empty plastic bottle in half, he used one half as a receptacle and the other as a funnel, placing a scrap of material in the neck to act as a filter.

Seeing the funnel inspired me to reply,

"No, not at all, the future is by its very nature multiple. Even without any free will, the future remains indeterminate, and even "indeterminist*". And that's not all: just think, the past could well be multiple, too. For you to create your mixture, there are lots of different ways of filling your funnel which all lead to the same result!"

"You're making my brain go all foggy!"

The expression was more apt than he probably intended. The mist from the summit was rapidly descending towards us. Soon we'd be right in the middle of a cloud...

Patrick's thinking was like that of most people blessed with a fairly good level of scientific knowledge. Most of the time, they think, we are simply conditioned to act a certain way, and our decisions are illusory, imposed on us by evolutionary necessity.

The Tree of Life I had attempted to explain to him the previous evening bothered him because of its premises: the authenticity of our free will on the one hand, and the hypothesis that our future already existed on the other.

Paradoxically, it was this last hypothesis that finally gained some agreement from him. It came from the fact that time is treated exactly like a dimension of space. Physicists and mathematicians have no difficulty conceiving this unusual way of treating time, because the Theory of Relativity has got them used to the idea already.

Albert Einstein himself, the father of general relativity, was the first to question the notion of time, and especially the notion of time in the present. The definition of such a present is entirely relative and subjective, because the very notion of a "state of the universe" at any given moment is completely false.

Observe a star in the sky, for example, many light years away from Earth. It is hard for us to imagine that this star belongs to the past. Yet it is quite possible that it has already exploded and disappeared—if it is a supernova, for instance. For us, this event belongs to the future, and so it does not yet exist, whereas as far as the star is concerned, the event is already in the past. The worst of it, though, is that Einstein showed us that objects do not even have to be far away in order not to share the same present with us: they merely have to move at different speeds. And this is why, whatever the zone of space may be, the present cannot exist in it!

Faced with the unreality of the present, the Tree of Life treats time like a dimension of space. This implies that we move through time just as we would in a vehicle travelling through space. This vehicle takes us to an unknown destination in time, but that destination is already present according to a multitude of possibilities, except if the vehicle is on automatic pilot, in which case we would have no free will: we would be pre-conditioned.

I had almost managed to convince Patrick about this question of time being assimilated to space. He was willing to believe that the future already existed, but in a unique form imposed on us by our pre-conditioning, whereas I was a partisan for multiple versions of it. And so it was the question of free will that caused the most debate between us. It was a much more arguable question than that of the omnipresence of the future.

The "premise" of free will is founded on the idea that at each new fork on our Tree of Life, we are free to choose one branch rather than another. This would make us truly free and responsible for our actions. We could only have a fixed destiny, a single and unique life scenario if our intentions always remained the same.

In spite of such a "chosen" destiny, our Tree of Life would still be relevant, because the slightest "authentically free" modification of our intentions would alter this destiny, and we would choose another branch. Our future would be controlled by our intentions: the probability of taking one branch rather than another would constantly vary according to our intentions.

Thus our future would be "indeterminist*", meaning undetermined by nature and therefore potentially multiple. Let us be precise about the difference between "indeterminist" and "indeterminate": an indeterminate future remains unique since it is only indeterminate out of ignorance of what renders it unique, as opposed to an indeterminist future, which by its very essence is multiple.

In the absence of authentic free will, we would be following a conditioned or "automatic" road already engraved upon our Tree of Life, such as not leaving the Road of Time, for example. If our conditioning was such that our life was programmed in advance like this, we may wonder whether our tree of life would still exist or not.

But wouldn't this be forgetting about the existence of chance*?

For chance could itself update* a new path along our Tree of Life, by making the choice for us. If we hit a wild boar on the road, for example, the accident could alter our route! And on a broader scale, chance encounters would replace illusory free will. Chance would make us branch off on our Tree of Life in one direction rather than another, choose this or that profession, or spouse, or home, rather than any other.

So if all our "choices" were really only imaginary ones, determined by chance or by our unconscious behavioural conditioning, then our Tree of Life would serve no purpose other than to describe a set of potentials ruled by chance alone.

And even then, chance itself would have to be the result of a true indeterminist process. Whereas, although the existence of this sort of chance is claimed by experts on quantum mechanics, the interpretation of results in this discipline is something still under debate. This is why, within the much larger community that includes all scientists, we still find that the dominant opinion is that chance is deterministic, but its causes are hidden: leaving no room for our Tree of Life.

Whatever the case, this materialist opinion of determinism* is what has dominated recent centuries ever since the Enlightenment, and still seems to dominate science even to this day. Determinism prevents a number of entirely respectable scientists from believing not only in indeterminist chance, but also in our authentic free will. It is especially true for most neuroscientists, accustomed as they are to carrying out daily tests on our brain—this fabulous piece of machinery with its multitude of neural interconnections: seeing that conscience is nothing but a product of the brain, they deduce that our freedom is illusory.

According to the majority of researchers in evolutionary, cognitive and behavioural science, our individual actions are determined by our preferences and by mechanisms which do not even reach our conscience itself (4), whatever their cause happens to be. And chance is itself a result of our ignorance of the causes! In a word, we are merely machines, in the same way that artificially constructed humanoids, equipped with human or superhuman intelligence, would be machines. We are not quite there yet, but technological progress makes it quite reasonable to suppose that we will know how to create such machines in the future.

Let us take the example of the first French humanoid robot, Nao. An association of experts from different research laboratories around the world, in which I took part in conceiving Nao's artificial eye, held the quite serious objective of making the robot evolve to a point where, by 2050, a team of humanoids would be able to win a football match against a team of humans! It goes without saying that this kind of goal, which completely ignores the usefulness of any "conscious function", only serves to reinforce the idea that consciousness can only be a mere by-product of the brain.

I have research colleagues who think that a sufficiently complex humanoid could magically acquire free will or even consciousness. Were they to be right, then we would have to give credit to the idea that perhaps we are no more than machines, exactly as determinism says we are.

Partisans of this powerful mechanistic paradigm* which has impregnated science since Newton, partisans who, until recently, formed a huge majority within the scientific world, thus consider that indeterminist chance does not exist, because its causes contain "hidden variables*" which act just like conditioning: chance, determinism or conditioning, it's all the same battle! In a word, our future is unique and already mapped out, because in theory it can be calculated, and we only have to wait for time to pass to find out what it is.

However, these ideas were called into question during the 20th century (23), first by quantum mechanics, with Heisenberg's principle of uncertainty, and then, much more recently, by Chaos Theory (33). Nevertheless, the Theory of Hidden Variables remains a powerful one, making a solid contribution to uphold the foundations of determinism which, along with causality, are actually the foundations of the whole of science itself:

– determinism: the future is determined mechanically and in a unique way;

– causality: the future is exclusively the consequence of the past, and not the reverse.

And in order to give good grounding to this last premise, physics has found nothing better than to hold up the principle of irreversibility—or the arrow of time—which tells us that it is impossible to recalculate the past from the present, because all movement is accompanied by increased disorder (or entropy) and so cannot happen backwards. And yet, all the laws of physics are reversible in relation to time, thus creating one of the profoundest mysteries of modern physics, and one that is still unresolved to this day.

In other words, some paths we take are irreversible, and to retrace our steps it would be impossible to take the same path we came on, going back over exactly the same points.

I was thinking this to myself as I clung to the rock I was climbing, when I heard Patrick say:

"Stop there, Philippe, or you won't be able to get back down again."

"But I wasn't intending to go back down again..."

"Careful, if you can't keep going up then you'll have to come down, and then we risk getting stuck."

"No, no, if we can get up then we'll manage to get down again."

"Absolutely not. Anyone would think you'd never done any mountaineering."

Patrick was right. In mountaineering, it is much easier to climb up than it is to climb down, quite simply because the irregularities of the rock face which serve as hand and footholds on the way up often become invisible on the way down. This means that coming down, one sometimes has to make long, perilous pauses between multiple hypothetical hand or footholds, and it is not unusual to fail to find the right one again. Mountaineering, then, possessed all the characteristics of an irreversible path.

During our walk up to the top of the Monges, the thick descending mist had induced us to rest at a place where a choice was forced upon us: either we continued walking through the mist along the narrow path, as we had originally planned, at the risk of getting lost, or we took a short cut by taking advantage of a cleft in the rocks we had found near our stopping point, and scaling a cliff. To save time, we had opted for this last solution, and were now half-way up the cliff.

"Anyway, I'm afraid to say it looks like we've no longer got any other choice: look behind you."

Keeping a firm hold with both hands, Patrick turned to look down and tried to see the route by which we had come up, but the cliff suddenly seemed so sheer that it made him pull back sharply, a wave of dizziness apparently sweeping over him.

"Good grief! We're right up the creek now. You got a rope in your bag?"

"No, but don't worry, we'll find the right route, come on, up we go."

As I said this, I was counting first and foremost on the psychological effect my confidence would have on Patrick. The moment for jokes was past, and instead it was time to exploit the supposed virtues of a determined and positive attitude. Even though the decision to carry on did indeed mean putting ourselves in the hands of a well-established determinism, that of a likely dead end that would only serve to confirm Patrick's fears, I wanted to believe in just one single, truly dependable determinism in the direction of the future: that of my entirely confident intent.

Patrick's intention, however, was even more convincing than my own, and so I jumped back down to his level and pondered the situation. Retracing our steps seemed impossible, as the climb back down really did look too perilous.

Could it be true that some journeys really are irreversible, blocking any possible way back?

In effect, certain physical phenomena seem to rebel against the fundamental principle of reversibility, and one branch of physics has made a speciality of it: according to statistical physics, time is irreversible. In fact, these phenomena are very common, and concern numerous physical systems, such as coffee machines, for example, or cars, motors, mixtures, cells, turbines, living organisms, etc.

It does seem impossible to envisage these systems working in reverse: a mixture will never transform itself by separating into its component parts, and a motor will never suck back its refluxes and function in reverse, etc.

On this point, however, the experts are still in doubt and divided amongst themselves, for the fundamental equations of physics are perfectly reversible: that is, they are symmetrical in relation to time. This is indeed the strangest contradiction in the whole of physics!

Having had the opportunity during my research career of creating models of irreversible systems, I had noticed that this contradiction was linked to the fact that such systems are by nature indeterminist. If we cannot calculate the exact position of each particle in a system that can send astronomical quantities of them into play, then all we can do is simulate movements of the whole by "randomly shooting" individual trajectories so we can calculate a global result of all their movements thanks to statistics.

The problem with this statistical method is that we cannot apply it backwards in time to calculate the state of the system as it stood originally, because statistics are incapable of finding what the initial conditions* were, that is, of reconstructing a past defined in advance!

Because of what I will term the "imposed" irreversibility produced by introducing chance into calculations, an irreversibility which, according to the scarcely more moderate opinions of numerous other physicists, is essentially engendered by statistics, some have wondered whether in fact irreversibility might not turn out to be no more than an illusion produced by a statistical effect (14).

However, this conclusion has kept a low profile, for it is hard to imagine how a bowl of café au lait could naturally evolve from a state in which the coffee is mixed with the milk into a state where the mixture has not yet occurred. And yet, theoretically it is possible, but only if we could wait for a length of time that would largely exceed the age of the universe!

This is why macroscopic physics, based on statistical equations, claim that it is impossible not only in fact, but also in principle, to "calculate" the past of phenomena said to be irreversible, and thereby reverse the course of time, which in this case translates into the impossibility of using a "calculation" to separate the coffee from the milk.

Whereas on the contrary, microscopic physics, in complete contradiction to this position, makes spectacular claims in its equations for the reversibility of all phenomena, and the possibility of making them work backwards and forwards in time!

So what is the truth? Here we have two totally contradictory types of physics: microscopic physics is complete and allows us to calculate, although only in theory, the future or the past from what we know of the present. But macroscopic physics, a statistical science which allows us to make entirely accurate and astonishingly effective calculations in practice, remains partial and empirical. It is incapable of calculating the past from the present and as a result qualifies certain transformations produced by time as impossible to reverse, on the pretext that there would be too much accompanying loss of information!

For the only thing physicists have come up with to justify irreversibility is the statement that "irreversible systems suffer loss of information"! Is this a reasonable conclusion? Do systems really lose information? Who really loses information?

Where had the information gone that was preventing Patrick and me from climbing back down the cliff the same way we had come up? Had it really vanished, or was it not rather that it was simply hidden from our view?

Seeing me leap down from my rock to join him, Patrick left me no time even to think about it, but launched into a downward climb that I found rash.

"Leave it, you won't manage it, it's too dangerous."

"I will, you'll see, there are only two difficult passages to overcome," he retorted.

"Wait, I'd rather we ate at the top. There has to be a solution."

"We've got time. Here, take my dried apricots."

To ease his descent, Patrick held out his rucksack to me to take some of the weight off his back. I could see that he was finding it hard, but he was so determined to climb down in spite of my own insistence on climbing up, that in the end I decided to help him: all I had to do from my position was lean out from a judicious spot and scan the rock face where he was desperately seeking footholds. That way, I managed perfectly well to guide him down by giving him the information he needed:

"Left, left, a bit more, yes, that's it, you've got it…"

Once he reached the bottom, I sent down our bags and undertook the descent myself, using his instructions to guide me. And that is how, by mutually giving each other information about where to place our feet, we managed to climb down onto a shoulder of rock.

If mountaineering allows us to fumble with our feet to find the right footholds and if necessary alter our route back down, in statistical physics it would seem that an arrow of time is trying to forbid us from such "fumbling", and even from making any sort of reverse journey at all to regain our initial position.

I use the term "forbid" advisedly, for in theory it is perfectly possible to calculate the past from the present. In practice, all one has to do is not force oneself to take the same route down, since we do not have access to the same information as on the way up. But this means we have to accept that alternative and even multiple pasts exist, and such an idea cannot help but shock determinist physicists, who are not even prepared to accept the existence of multiple futures!

This being so, does it mean that forbidding us to reverse the arrow of time is tantamount to forbidding us from returning to a different past? In which case, **would it not be more judicious to talk about inverted determinism?**

Yet surely there was a quasi-infinity of ways for Patrick to pour the water into his funnel to concoct our drink. After all, if the concept of the tree of life is correct and physics is indeterminist, then it has to be so in both directions of time, because its equations are reversible.

It is important, then, to give this some closer thought: when we question the notion of irreversibility, it unfortunately leads us, at first glance, to admit the existence of a multiple past! How can this be possible?

What we will do now is show that this is simply a technical problem, and just as statistical physics is merely a calculation of the future from the present, the fact that we do not know how to calculate the past from this same present does not actually mean that it has to be multiple.

First of all, let us note that the property of Patrick's funnel is not at all a typical example of going back in time—quite the opposite in fact! Because exactly the same thing happens in the future of irreversible systems, meteorological phenomena for example, or the rubbing of a wheel, or the combustion engine: the particles of these systems disappear in totally unpredictable directions.

However, we have noticed that in spite of their apparent calculability, many irreversible systems become entirely unpredictable in the short term, whether they are living or artificial systems, because the calculation swiftly becomes completely false. Any calculated prediction of a future state becomes impossible to carry out as of a certain point in time. Today this observation has been effectively explained by Chaos Theory (24), and the best illustration we know of can be seen in the limitations of weather forecasts.

As it happens, this observation of unpredictability is the same as the one which prevents us, when going back in time, from reconstructing a unique past! And yet no one says that going back to the past is irreversible, because that would be utter nonsense!

This is how the popularity of Chaos Theory, like that of quantum mechanics, has led physicists to acknowledge once and for all the "natural" unpredictability of the future. For even if we did have gigantic calculators, there would still be a temporal limit to prediction, due to a sensitivity to initial conditions* right down at microscopic level: the level where a certain fundamental indeterminism already reigns, concerning the position and speed of elementary particles. It is impossible to have precise knowledge of the one without the other being indeterminist.

It follows that present-day physics contradicts the mechanistic paradigm (causality + determinism) upon which it is constructed, by making indeterminism reign at every level: a more precise argument for this fundamental statement in favour of Double Causality Theory can be found in the annexe at the end of this book.

Faced with nature's general indeterminism and the reversibility of equations in physics, then the past should be indeterminist as well!

But would that not be completely to ignore memory of the past? What do we do about the information available to us? What about the mutual information strategy that Patrick and I adopted to find our way back down?

Now we finally have a good reason for questioning the notion of irreversibility that results from statistical physics: the problem with statistical calculations is that they do not let us "fumble" to take account of the slightest trace of the future or the past, since this is an actual technical impossibility, especially as all such calculations are based upon the liberal use of chance!

As we have seen, this notion of irreversibility has managed to prevail by claiming so-called loss of information, corresponding to an inevitable increase in entropy*. When we take a closer look, however, we see that this information is not lost at all, providing we are prepared to make use of all traces of the past, such as the open litre of milk standing next to the bowl of café au lait, or even the coffee grounds! The problem is that we simply do not have the tools for taking these traces of the past into account. The fact that the information has been completely diluted or dispersed does not imply that it has actually disappeared.

At worst, we would only have to find the person who had made the mixture, and their simple statement would suffice to reconstruct the bowl's past. Not that this would serve any purpose, because that person is already part of the "established" observers* of the universe: their brain contains a trace of the bowl's past!

An irreversible system's information is only lost by a physicist who mistakes himself for a technician by observing the technical impossibility of finding that information and making the reverse calculation, but the universe has never lost that information!

Consequently, even if in theory the past can be calculated because of the reversibility of general equations, we do not know how to calculate that past, because we simply do not have the tools to do so. We only know how to make qualitative reconstructions which, incidentally, is almost sufficient in itself to call irreversibility into question.

Are we missing some equations? We obviously lack a sort of "inverted statistical physics", which would actually be much less statistical, for it would have to be able to exploit traces of the past to find the initial conditions. Just as present-day statistical physics would need to know how to exploit the "final conditions" or "traces of the future" in order to identify the different possible futures of a single chaotic system.

The fact that the past is incalculable does not therefore mean that it is indeterminist, but that we simply have not yet found the law of "effect to cause" which allows us to work backwards to a unique past by bringing together all the traces of this past.

We would have to accept, therefore, a sort of "inverted causality" or retro-causality*. This concept has already been put forward in quantum mechanics, as well as to explain certain parapsychological phenomena (25), but physicists dispute it because it goes against the standard model of physics.

Retro-causality would stop being questionable, however, if it were accompanied by the acknowledgement of inverted determinism. This is why I prefer to use the term "second causality" to qualify it, making it clear that we are dealing with "determinist retro-causality". Another good reason to make this distinction is that retro-causality usually considers that the past precedes the future, which is not the case with second causality. This latter is in fact based upon the omnipresence of the future and the past, on their "simultaneousness".

Second causality, then, interprets irreversibility as a technical problem: on the one hand, we do not know how to exploit the information contained in the traces of the past we can observe in the present. On the other hand, we do not know the law of information to substitute for statistical physics in order to calculate the past from these traces.

In no way, then, are we dealing with an indeterminist past!

Quite the contrary, in fact: since traces of the past should be much more numerous than traces of the future, then, based upon this simple fact, we can state that the past is much more determinist than the future, and if we were to admit that we had perfect knowledge of our past, we could make the following assertion:

Determinism is only valid if it is inverted.

Let us now see how this observation holds up in relation to our Tree of Life: inverting determinism corresponds precisely to the fact that if we climb down our tree, we invariably get back to the trunk. Inverted determinism, then, is based upon the fact that traces of the past render it unique, as opposed to the future, whose traces should be much less numerous, or even non-existent.

But what about this law of information which would be able to recalculate the past from traces of it?

Let us remember that this law would have to be reversible, in order to satisfy the fundamental equations of physics! So it would have to encompass statistical physics and reach the same results in the particular case when no traces are brought into the calculation!

Physics, then, must be incomplete, in that we only know how to calculate how a system is going to evolve in the absence of any traces, by choosing the most probable scenario. But what if this was simply another technical problem?

Surely the fact that the result as calculated by statistics complies with our observations signifies rather that we are measuring the most probable result?

But what if sometimes the traces of a particular future were well-defined, thus favouring a rare evolutionary scenario—should we not then expect a result which complies very little with the calculation of probabilities?

We will set off in search of these traces. Whatever the case may be, let us keep in mind that by inverting the direction of determinism, we have also put forward the very plausible hypothesis of the indeterminism of the future. This fundamental indeterminism of nature, for which I argue at the end of the book, will give second causality the possibility of exploiting all the richness of a considerable amount of chance events*!

Chapter 2—Summary

We must differentiate between two types of choice: conditioned—or determinist—choices and choices which are genuinely free or indeterminist. The Tree of Life symbolizes the latter when we are climbing, but also demonstrates the illusory character of the former when we come back down.

This supposes an inverted determinism, which has forced us to reexamine the notion of irreversibility, by considering it as a technical inability to recalculate the past from the present, due to a loss of information that is entirely relative.

In reality, if we knew how to use all the information or traces of the past that we can observe in the present, and knew the law allowing us to piece them together, we would see that the past is much more determinist than the future. Inversely, it is the apparent dearth of traces of the future that makes it indeterminist.

III.

THE LAW OF CONVERGING PARTS

*In which we discover that the search for the missing law
for calculating the past leads us on a quest for traces of the future.*

To make it easier on the way down, we took a different route to the one we had taken coming up, and a pleasant surprise awaited us at the bottom: perched on a cliff shelf was a magnificent slab of fossils forming a superb natural table, just flat enough to host our picnic. As we ate our lunch, the mist cleared and completely dissipated above the summit of the Monges Hills.

"What excellent intuition of yours to decide to come down this way!"

"No, not at all, we'd never have got here without your help. I was that close to following you, all you'd have had to do was reach out your hand."

"Well, maybe I had some good intuition too... Hmm... I do believe that if we'd gone up, we would have got stuck."

"Look at the shape of that one! It looks like a sea-horse!"

"And when you think that it's hundreds of millions of years old!"

"Look over there, next to it, it looks like they were fossilized at the same time. This slab is small, but much finer than the one at Fontbelle, and probably nobody knows it's here."

"Have you seen where we are? That's hardly surprising. Besides, I wonder whether it wasn't formed quite recently. Look at that enormous rock, it looks like it's only just fallen."

"Phew, you're right! It's not the only one, either, I'd say all the others nearby came from up there, too. Better not hang around here too long!"

"I expect they were brought down by the recent bad weather. If you ask me, that one broke off about ten days ago when we had that spell of hard frost. And look here, this piece fell on the tracks of a mouflon."

"Oh, then it must be much more recent than that!"

"No, it hasn't rained since, and it's dried. See? There's dried mud on that one, and yet it seems to have fallen from up there too. The mouflon must have splashed mud on it as it trotted past, before the rockfall stabilized."

"Well, well! You'd make a splendid Sherlock Holmes!"

"Oh no, none of it's certain, you know. And anyway, all these traces of the past are surely more uncertain than the trace of the future we followed before we found the cleft in the rock!"

"What do you mean? What trace?"

"Don't you remember?"

Long before Patrick had decided to pause at the foot of the cliff and wait for the mist to clear a little, we had reached a point where we could not make out the footpath. Even if we retraced our steps, we could not be sure of going the right way. It had seemed more sensible to go back home. But suddenly, as we stood there hesitating, a ray of sunlight filtered through the air for a few seconds, like a promise to light our way forward, and we looked at each other and Patrick said, with a determined air,

"Right, on we go!"

And indeed, an hour later, at the end of our eventful climb, the sky had cleared.

Patrick's wise decision seemed to have come from a ray of sunshine. Could it be a trace of the future? To find that out, perhaps we need to know the law of information that allows the present to be deduced from the future. To do that, however, surely it would be wiser first to look at how to reconstruct the past using reliable traces.

Could there be a methodical way to bring a Sherlock Holmes approach to the work, and make the job of reconstructing the past easier?

Let us go back to the example of the phenomenon that appears to be the most difficult to reverse, that of a mixture between two constituent parts, like café au lait: we can always find traces of the mixture's initial state, and we can find them whether they be internal, like unmixed micro-residues, or external, like the packet of coffee or the carton of milk near the bowl, not to mention the memory of the person who has had his breakfast.

This is how the hypothetical loss of information about a system through degradation or mixing may well be an apparent loss only, one that could never prevent us from reconstructing its past state, because of the massive redundancy of traces! For example, to reconstruct a broken glass from hundreds of shards scattered on the floor, we do not need all the pieces: to deduce the shape and position of the original glass, just a few will suffice. Conversely, statistical physics would be completely incapable of calculating the shape and position of so many scattered pieces!

It is the same for a car that has been in an accident: it is much easier to reconstruct an accident after it has happened than to describe it before it happens, because most accidents involve redundant human, mechanical, meteorological and local factors whose convergence on the time and place of the accident is entirely unpredictable.

The reader familiar with the concept of entropy* will notice that I have taken examples where entropy increases significantly. This is a generality, for in the sense of time, entropy always increases.

We see then that the direction of time is much more inclined towards all forms of dispersion, mixture, dissolving and degradation than the reverse direction, and that is why the future is so unpredictable. Going back in time, however, is favourable to reuniting different parts, so much so that it happens automatically according to a qualitative law of convergence which we could explain thus:

– everything that bears a relationship in spite of distance converges together in the direction of the past to form an ordered system, or, in a more concise, albeit less precise, version:

– like calls to like.

Could the law that manufactures order act as a law of "effect to cause" and so complete statistical physics in reconstructing the past? We would still need to study the way it operates in order to make it quantitative and thus deduce how to formalize such a law.

First of all, it would need to maintain energy, meaning that energy would have to be "sucked" out of the surrounding environment, so that a ball could bounce higher and higher and finally return into the hands of the person who had thrown it, for example. This is something we have never seen happen, but nothing about this process of aspiration through "converging parts" would contradict the reversible laws of mechanics, except for the infinitely small probability of it happening without imposing the initial conditions for it.

Conversely, physicists are already hard pressed to impose any particular final condition in the presence of dispersive phenomena, whether they be chaotic or turbulent in nature: so the problem exists in both senses of time. Even when they are sure of the possibility of one outcome among a number of others, they cannot deduce the evolutionary path which leads to this outcome. To exaggerate a little, it would be like trying to make a calculation model specify whether we would prefer the action of a butterfly's wings* in Australia to provoke a storm in the Caribbean rather than in the Mediterranean!

Or to put it another way, it would be like asking the "author" of our hiking scenario whether he would prefer to find us stuck up a cliff waiting for the helicopter to come, or see us succeed in reaching the top by some unsuspected route.

For finally, after our improvised feast on the fossil slab, Patrick and I quickly picked out a new route for climbing back up the cliff, by following a second cleft which took us to the top in no time at all: "chance" appeared to smile on us as the sun came back out. Patrick was astonished by this reversal of events.

"I don't believe it. Climbing down using each other as guide has brought us some incredible luck. I'd never have thought things could have turned out so well!"

"You're the one who made the right decision, to climb back down."

"No, you're the one who decided to help me do it."

"In that case, you know what? We were standing at a "fork", and our future drew us towards it. All we did was look at both possibilities in a detached enough way. That way we were ready *to choose* spontaneously the one which was leading us to that future."

"Ha! Ha! Ha!... That doesn't surprise me, coming from you, Philippe, but no, come on! We were lucky, that's all, and we only ended up finding this slab by chance..."

By chance? Statistical physics has actually applied an empirical law to itself that uses chance to predict the future, but it fails entirely when we try to force the future to adopt certain states, and I would even go further and say: these states are "creative", and entirely possible.

"If you ask me, your chance wouldn't even have helped us to find the first cleft in the rock, and we would have got lost at the foot of the cliff", I answered Patrick.

In physics, knowing how to list and calculate all the different possible outcomes for a dispersive system to the point of being indeterminist, even if this means assigning mere probabilities to them, corresponds to a real need. Today, we have trouble calculating the multiple possible evolutions of such a system. In my laboratory, however, by concentrating only on stable or balanced evolutions, we are beginning to be able to do this, and it is a research subject I share with certain fellow professors at the University of Provence.

For example, when we cause fluids to circulate in a tube or container using thermal effects, we obtain different structures such as oval swirls or hexagonal cells, which may become too unstable to allow the precise regulation of certain parameters. We then make digital simulations use calculation programmes which allow us to find the exact structure of these shapes. We have noticed that within these conditions of instability, the structures we calculate can sometimes be very different without actually having changed anything about the initial digital conditions, even when working with the greatest amount of data precision. Everything depends on whether we went off to get a coffee at some point, or on the computer we were using...

In order to understand this variability and find out whether this is indeterminism or whether we are in the presence of hidden variables, we would be hard pushed to start from our final state and find our original conditions in order to evaluate them, for the state has become stable and dissipated the original information. Even were we to memorize it when it was still in a state of disequilibrium, the calculations would not allow us to find the original conditions, because we do not know how to achieve the reverse process of dissipating energy information. To achieve such a process, we would need to be able to synchronize all the trajectories corresponding to this dissipation, including the radiative dispersion of photons, using information we do not have.

For this, we would also need a Law of Converging Parts* that would use the information allowing all these trajectories to be synchronized while making them converge on a predetermined past. Logically, the information should originate from the initial conditions, or even from any other traces of the past, which would "aspirate" everything it had dispersed. Nonetheless, this law should be intrinsically determinist by the very fact of its principle of reconstructing synchronous order.

This, however, is a hypothetical missing law, and all we can hope for here is a qualitative description of it, since we do not know the equations for it.

Let us see whether we can try to gain a qualitative understanding of this law's determinist reconstruction process. And let us begin by remarking that it is much easier to do this by going backwards in time than by going forwards!

Indeed, it is much easier to predict nature's past from knowledge of its present than the reverse. This is true notably on a geological scale: geologists have fairly precise knowledge of the events which happened on earth hundreds of millions of years ago, and know how to date them with precision, simply by observing and analysing the geological strata we can observe in the present, and they can do this even though extremely complicated folds of mountain ranges have been formed. Whereas in the opposite direction, it is impossible for a geologist to predict the relief our landscape will display several hundred million years hence, except by making unrealistic simplifications.

Let us note here that this conclusion has been discovered only recently, as scientists realized that chaotic behaviour could even affect the apparently most predictable systems, such as our solar system, and that this meant we had to revise all our simulations in the direction of time. And I will not even mention the Big Bang*, the origins of our universe, something physicists today can describe with breathtaking quantitative details (28), whereas we are still just as incapable of knowing what might happen in a length of time a thousand times shorter: the end of our sun? Of our planet? Or simply of our civilization? And if so, when and how?

It is in fact much easier to predict the origins of our universe by making all the information we possess about the past converge together towards its ultimate state: just as all water flows to the sea, we know that all the stars in our universe originally came out of a "primaeval soup" which, as it gets smaller and smaller the further it goes back in time, constitutes a mass of matter under more and more pressure, moving back towards the hyper-homogenous state of the universe at the time of the Big Bang. And we would not even think of forgetting a star or galaxy in our reconstruction of the Big Bang, it being implicitly understood that all elements of the universe must be in it.

Going back in time to a homogenous Big Bang surely constitutes the best possible illustration for showing that there is a Law of Converging Parts at work in the universe.

What does this law mean if we put a human being into the equation? Here, too, we can verify that simply having knowledge of the present makes it much easier to predict the backwards evolution of someone than his normal evolution, when we reverse time.

When in fact we reverse the "direction of life", we notice that we do not observe the same incertitude at all about an individual's "future" by "playing out" his lifetime from the future towards the past. If you need convincing, you only have to consult an adult in good health to have an idea of his remaining lifespan before he returns to the cradle: quite simply his current age! And also to be in absolutely no doubt that during the whole of that time, he will live a life free of major accidents until the moment he returns to an embryonic state in his mother's womb, and, moreover, to know in advance who his mother will be, along with an infinite number of other details. It is even clearer than with non-organic matter, for it is impossible to disperse a living being's matter in the past without risking being unable to reconstruct it entirely in the future afterwards, something that is a compulsory part of reversibility. So it is hard to see how we could propose alternative versions to the final stage of the embryonic state and of its DNA.

Once again, surely this is the best possible illustration of a synchronous order system working in the opposite sense to time, using strands of DNA to make a meticulous arrangement of all the information which, once unravelled, will allow a human being to be created?

Thus we observe that one of the fundamental characteristics of the backward progression of time, or a return to the past, is to create order that reassembles, in a minutely synchronized manner, everything that allows life to be created.

In the final part of this work we will see that the Law of Converging Parts that allows the creation of this synchronous order is nothing other than a **law of attraction of temporal lines** or, if we consider our destiny, of our **"life trajectories"**, and that it is not unrelated to universal gravitation. Let us not forget that the past and the future are omnipresent and simultaneous, although it is too early to develop that point quite yet.

For now, we have managed to identify why there is an arrow of time which makes "living in the direction of the past" very different from "living in the direction of the future", despite the reversibility of time according to physics: it is because there seems to be no "trace of our future" left on our current present, something which would otherwise let us live with a sort of "memory of the future"!

However, if we look closely, we gradually realize that our future leaves many more "traces of the future" on our present than we could possibly imagine, but the problem is that we are not used to considering them as such. Soon we will be discovering subtle manifestations of such traces, but we can already ask ourselves whether some clues are not simply that: you only have to look at a young child expressing inborn talents, and watch him talk, play or have fun, to get a fairly precise idea of what will mark him out from others in the future. This poses the question whether an obvious vocation or inborn talent should be considered a trace of a particularly creative future life?

Our answer to that will be **no**, unless we accept that a single trace might be both a trace of the future and a trace of the past, something we are not going to do. So we will consider the newborn child's vocation as the inheritance of a fortunate genetic combination. This will lead us to observe that the reversibility of time prevents us from easily detecting exclusive traces of the future: the very ones that interest us!

So we have already gathered at least two reasons to explain why traces of the future should be so difficult to find. The first is the indeterminism of the future, in utter contrast to the determinism of the past; and the second is the fact that we exclude from traces of the future any which are also traces of the past!

Hardly surprising, then, that our future memory should be so poor, over-encumbered as it is with causal explanations!

Not only that: add to this our supposed free will and now we have three explanations for our having so little "future memory", the third one being that were we to possess such memory, we would be tempted to modify it using our free will, all the more so if it contained suffering yet to come (even if we did not regret it once we had lived through it).

If we want to hope to find traces of our future, then, we should seek them among futures we have no intention of modifying, and which are not caused by our past! Our free will absolutely has to come into play, for without it these traces of the future would be no such thing, because they would be explained by the past, in other words, by causal means. So we have to be free to modify this future, but without actually doing so! But what would be the point of liberty if we simply had to accept our destiny? Is this a squaring of the circle, or a puzzle set by some Buddhist sage or Cartesian monk? All I can say is that as I see it, this is one reason why the existence of these traces remains doubtful at least.

In the end, surely the absence of traces of our future is simply a reflection of our absence of free will? The question deserves to be asked.

Whatever the case, if our free will really is authentic, then we must look for traces of our future somewhere it can act upon or even create such traces! And what does our free will need in order to act, if not our intentions? Whatever we do freely, all of it is in fact predetermined by our intentions.

However, let us not go too fast; let us return to the Road of Time to get a better look, for all this analysis can be reduced to a single metaphor: it is much easier to come back down a mountain and unfailingly get back to the main road, even if we take different paths, than it is to head for a completely unknown destination of which we have no trace: in the first place we are conditioned by the descent, and in the second we are free.

Numerous questions then arise: what is this freedom? Is it a quest? If the quest is that for Theopolis, should we not be able to find traces of it? How can we recognize such traces? How can we detect these traces of the future we have freely chosen on the Road of Time? Do we not need a better idea of such a future, in other words, a clear intention? Armed with an intention we suspect will create traces, what else do we need in order to find them? How can we finally learn to distinguish the traces of a future that actually concerns us, now that we are more resolved than ever to liberate ourselves from our conditioning and develop our own free will?

On the Road of Time, it is obvious that our choices are made at forks in the path, but we still have to detect the presence of such forks, for there are paths, even the faintest tracks, all over the place. Observation, then, must have an important role to play, as important a role as intention, in fact, for without observation we would not even be aware of a potential fork in the path before us, and so we would keep within a determinist scenario.

However, if we observed a trace of our future suggesting we should follow another path, what information would we use as the basis for choosing to branch off? And how can we be sure that our own free will remains entirely free in making this choice?

Whatever the case, the role of our ability to observe and pay attention is evidently decisive: intention and observation, then, form the two wings of our free will!

Let us take a little detour via modern physics to see what it has to say about observation: it is tempting to compare the importance of the observer's role in exercising his free will to that of the role of the observer in quantum mechanics. For one of the most recent leaps of progress in quantum physics provides fairly precise information about this privileged role.

We are talking about decoherence theory. The concept of decoherence describes how the simultaneous coexistence of several potential "branches of life" manages to be sustained before reaching the ultimate stage of observation, through a clever process of "decorrelation". In reality, it is a question of particle trajectories, but even physicists allow themselves to make this sort of extrapolation.

Decoherence explains the mechanism of how a unique observation, from among a whole set of possibilities, enters reality. Even though these are only "microscopic branches", the observation deserves to be looked at, because it leads to the emergence of a single version of reality, "one of many being prepared".

However, as our present knowledge stands, the ultimate observation process that lets us bring a "potential branch of life" (of a particle) into our reality in this way involves a choice which, for want of anything better, is attributed to chance. But could it be attributed to the future?

Knowing as we do that it is reasonable to doubt choices made by chance, something Einstein himself repudiated, refusing to believe that God played dice, and knowing that such chance produces an existential vacuum that denies our free will and bears little resemblance to the concept the human race may have of its role in the universe, we can reasonably agree that the process of how one of our potential futures enters our reality is a phenomenon of which we have little understanding, according to the present state of scientific knowledge. Might it not be because we attribute this choice to the present?

We will go further into this question relating to traces of the future, by providing an interpretation of the quantum phenomenon of decoherence, something that finally operates in favour of our free will.

For now, though, fortified by the idea that our free will exists, that it depends on the spreading of the two wings of intention and observation, and that the Law of Converging Parts may well offer traces of our future to our process of observation, we will now examine how this process, rendered so mysterious by science itself, could split time into two parts:
- a time to live our pre-traced destiny;
- a time to change our destiny.

Chapter 3—Summary

To employ inverted determinism, we need a missing law capable of creating synchronous order, a law we will provisionally label the "Law of Converging Parts".

This law is reversible and consists in calculating the past or the future from the respective traces of each. When we employ it in the absence of any traces, it becomes physical statistics again, using chance to fill in the gaps left by the indeterminism caused by the absence of traces.

No traces of our future can exist except those which reflect our own free will.

In order to produce such traces, we must inevitably confront the authenticity of our intentions.

Our ability to observe these traces is fundamental in detecting them, for it is what allows us to discover every possible fork in the path on our journey along the Tree of Life.

And yet the fundamental importance of the act of observation in the manifestation of a unique reality is also a product of modern physics.

IV.

RIFT IN TIME

*In which we look for the characteristics that traces of the future
should reveal to our observation, and discover a first rift in time.*

One day, as I rounded a chaotic path in the geological reserve, my eye
was caught by a roe deer almost colliding with me, crossing my path at high
speed and heading for a rocky ridge leading to a sheer drop that offered him
no way out. I had been following the path's rather patchy traces for a good
hour without managing to get off the endless ridge it ran along. As time
passed, I moved further and further away from my goal, which was to climb
the summit opposite. So I decided to slip into the animal's wake, managing
to take a photo of it just before it disappeared, and I swiftly came across
an unexpected cleft in the rock. Perhaps I still had time to reach my goal? I
realized that I did not, for the cleft was overgrown with thorns, hindering my
view and the speed of my descent, and I even began to wonder whether I
would be able to get back before nightfall. With much effort, I had managed
to get about half-way down when a company of partridges flew up and
caused me to look around. Suddenly, just for a second, I caught sight of
my roe deer standing on a pile of scree, before vanishing behind a slope.
My intuition told me I was on the wrong path, and that the route I had been
seeking for nearly an hour must be somewhere beyond that slope. I could
not say why, but even though I could not seem to find any rational foundation
for this feeling, I decided to listen to it and retrace my steps. After managing,
with great difficulty, to extract myself from the cleft via a barely discernible
exit, it did not take me long to establish, as I succeeded in reaching the
pile of scree, that I had been right to trust the roe deer: looking back up at
the way I had gone, my initial route would have been blocked by another
unclimbable cliff that I would have had to walk along for a long time before
being able to continue my descent. And night would have fallen before I
could have found the right path!

It was not until much later, months in fact, recalling this walk and how I
nearly got lost as night fell, that I realized my intuition had been grounded
on that sudden flight of partridges!

I would not have attached any importance to the coincidence between the partridges and my fleeting glimpse of the roe deer, had I not at that moment been busy pondering the notion that coincidences might perhaps be explained by the Law of Converging Parts. This made me suspect that they were traces of the future, and I wondered if, as a general rule, our intuitions were something approaching an almost unconscious detection of these traces of the future, in the wake of one or more observations.

Let us look at this Law before studying how it behaves in the presence of chance.

In order to represent our life by an image that preserves our free will, we have introduced the model of the Tree of Life. This has led us to question the sense of traditional determinism, the very thing that wants us to believe we have an unshakeable destiny.

We have seen two interesting aspects to such an inversion:

1. it offers a solution to the incompatibility between determinism and free will by inverting the former;

2. it suggests the existence of a Law Converging Parts as an alternative to the Law of Increasing Entropy.

Two things justified us offering this sort of alternative. Firstly, the failure of many physicists to be satisfied by the ban on reversing the arrow of time, which contradicts the fundamental laws of physics; and secondly, the incomplete nature of "statistical" physics, which, as the adjective suggests, is powerless to find anything better for predicting the future than the exaggerated use of chance and probabilities, even though this does, from a mathematical point of view, represent remarkable prowess.

Even if it only serves to move away from the exaggerated use of chance in our calculations, which lies at the source of the conclusion that entropy and therefore disorder can only increase indefinitely, it is interesting to study how our alternative law finally rids us of chance and these sorts of beliefs.

First of all, though, could the Law of Converging Parts get rid of the element of chance in my coincidence with the roe deer and the partridges, by proposing a "future cause"? How do we arrive at such an explanation? Do we imagine the potential future of these animals, or merely the effect the coincidence may have had on my own future, on my Tree of Life?

One thing is sure: without my presence, there would not even have been a coincidence. This means I am firmly implicated in the matter. So let us us consider my own future: I escaped the ordeal of a night spent out in the open, in the wild, with no torch or warm clothing, lost somewhere in the mountains where, even by day, it is hard to find one's way.

Was there no scenario in the memory of my Tree of Life which had me falling ill after catching cold? Or did it have another compulsory scenario to impose upon me at this moment in my life? Let us admit that this does not seem to stand up, and moreover offers no mechanical explanation for why the partridges flew up at the precise moment the roe deer reappeared fleetingly in my field of vision. In fact, the explanation seems to be quite simple and purely causal: the partridges flew up because I came along, and I was lucky to reach their level, quite by chance, just as the roe deer became visible again: more chance.

But this does not stand up, for it is precisely the low probability of such a chance that we find so implausible, and anyway, it is reasonable to think that I had come to a fork on the path of my life, in other words, that I was faced with two branches: I could choose to follow the roe deer or I could choose not to do so, for I believe that, consciously or unconsciously, my free will was responsible for my decision, against all rational expectation, to follow on the trail of the roe deer.

I will take the liberty of placing partridges and roe deer in the category of uncertain "traces of the future"—my own future, as it happens—without actually explaining them for now. For that, other developments are first necessary.

We can well imagine that a fairly high level of uncertainty reigns over traces of the future we might be able to "glean" this way, even more so than with traces of the past. They may be quite distinct in nature, whether we are dealing with sophisticated calculations or coincidences, the consulting of oracles or of anything we might generally qualify as being "signs of destiny".

What is of interest to us here is not knowing whether what we suspect to be traces are related to certainties in the future, but learning how to distinguish them from false traces, or, in other words, from illusions, predictions and all kinds of silliness.

For even if they are uncertain, good traces are still worthy of interest.

The missing Law of Converging Parts gives us the following clues: by applying its mechanisms for creating synchronous order in the direction of the past, we can first check that the traces of the future imply observation of a group of elements that is much more ordered in its present than in its future state, in accordance with the rule for creating order. However, we must exclude groups where such order already exists or would be even more ordered in the past, as is the case for any system in the process of breaking down, for example.

This means that a trace of the future can only appear in a synchronous, ordered form if the order has no causal explanation, because otherwise such order would merely be a trace of a more ordered past.

Looking at it this way, can we consider my animals as a trace of the future?

There is indeed creation of synchronous order here: the harmony between the flight of partridges, my head turning, the appearance of the roe deer, not to mention how much it was in my interest that all this should happen at the same time and in such a coordinated manner. The least delay, the least lack of coordination, would have created disorder by making me feel I had missed the chance offered me of escaping my fate of sleeping under the stars and catching the nasty bout of bronchitis that would have ensued.

In general, any coincidences that are difficult to put down to chance, or, in other words, any coincidences for which a causal explanation is highly improbable, can therefore be suspected of being traces of the future, the order they create arising from the simultaneous or proximal nature of the elements comprising them. Improbability intervenes as a measure of the perfection of the synchronism, whether its nature be temporal (converging at the same moment) or spatial (converging on the same place).

It seems, then, that what really characterizes this improbability is first and foremost its synchronism. As far as order is concerned, we would first have to understand the meaning behind the coincidence, if indeed there is a meaning.

Coincidences, therefore, have something to teach us about a potential property of the Law of Converging Parts, which might provide a non-causal explanation for the creation of unexplained synchronism!

Yet it is easy to understand how, in order to come back from a starting point in the future where I am happily asleep in my bed by evening, to a past where I have lost my way on a path highly likely to make me spend the night out in the open, it is indispensable to bring about the convergence of a set of elements from the environment that are linked to each other by their ability to unite the synchronous conditions resulting in my finding myself at the top of the cleft, at the spot where I took my photo of the roe deer.

So what the Law of Converging Parts had to do was reconstruct a past of which this photo was one trace, by finding some way to make me discover the almost imperceptible exit without which I would have spent the night out of doors. Have you ever tried to follow a roe deer? It was quite impossible for it not to disappear from my field of vision! Clearly, the Law of Converging Parts had little other choice for bringing me to the top of this mountain ridge, and it did all it could in coherence with **the game contingencies* of the surroundings**.

It seems, then, that this law can lend itself to inverted causal reasoning, thanks to the indeterminism of nature, thus acting as a competitor to causality, and this may shock the sensibilities of some. At this point, we will suggest only the hypothesis that the idea of causality may, due to this indeterminism, be complemented by competing, apparently non-causal mechanisms. This, quite simply, is the key to our search for traces of the future. It would be good to get used to the idea of "non-causal logic" right away, to the idea of an inverted causality that is nonetheless perfectly rational.

Let us now see how all suspected traces of the future are usually treated by causal logic: within the framework of the present paradigm of our conception of time, we tend to consider all such traces as being unfounded, whether they are signs, coincidences, predictions, clairvoyance or other sorts of spells. When they are labelled as signs, then it is put down to personal intuition, something which is tolerated to a certain extent, but as for the rest, causality only tolerates predictions based on determinist calculations.

Popular intuition itself, faced with traces of the future often labelled as "signs of destiny", rarely considers them reliable, and puts them in the category of dubious visions or mental projections, sometimes unhealthy ones, and in doing so risks losing a certain amount of information.

This refusal of any non-causal explanation is also what disturbs physicists confronted by the reversibility of the equations of physics, and does indeed constitute a challenge to causality. The only thing still stopping us from looking for traces of the future is the dogma of irreversibility. According to that dogma, we are condemned to suffer the ensuing disorder and unpredictability, and must float in an ocean of uncertainty.

The major problem with this dogma is that the equations that help establish it work in both directions of time! Which means that the past should also be more disordered than the present and that going back into the past should also be irreversible.

Would you accept that an official investigator, looking for traces of the past in the course of a legal procedure, for example, should systematically be incapable of ever offering the least certainty about his conclusions, on the pretext that they would always be susceptible to change, as and when the traces disappeared, for instance? No, you would not, because you would judge that in the majority of cases, acquiring certainties is all part of an investigator's job. Conversely, do you believe your future might never be determined, even when your firm intention actually allows you to ensure it? Surely it would be too pessimistic to accept a law of increasing entropy that, on principle, would remove any possibility for having confidence in life, keeping us in a permanent state of fear that everything could go wrong at any moment?

Causal logic, foundering on the dogma of the arrow of time, sends us back to face the avatars of irreversibility, unpredictability and dispersion, or in other words, a pure, existential vacuum.

On the contrary, with a Law of Converging Parts that still needs to be formalized, we are justified in hoping to be finally rid of excessive uncertainty since, as opposed to an evolution towards the past which is independent of our will because it is determinist, the evolution directing us towards our future seems to confront us with real choices.

We still need to know how to exercise such choices and maintain the intentions attached to them.

We can see that the primary source of certainty concerning any traces of our own future we may hope to encounter, according to our Tree of Life, will necessarily bring into play our intentions, and therefore our own integrity or authenticity as regards our ability not to lie to ourselves. If we want to observe traces of our own future, then the least we can do is seriously examine our intentions, if only so as to avoid exposing ourselves to contradictory traces.

The condition that our intentions must be authentic faces us with a contradiction: if they are to be reliable, it is preferable our intentions be accompanied by the means to achieve them.

But if that is the case, then the achievement ends up being attributed to causality, so we can no longer hope to see any trace of the future of such an achievement! Now there's a dilemma for you!

This would explain why we see few, if any, traces of our future, quite simply because by imagining the means we might use to achieve our objective, we inscribe our future potential acts into a causal strategy, one of these means thus becoming the cause of our future. And this, then, destroys any possibility that that particular future be derived from anything else but these means, thus making it impossible for any non-causal traces to appear!

Therefore, a third condition has to be added to our search for traces of our future: **we should be determined, but not yet know how to realize our intentions!**

So there is a potential conflict between the two opposite types of traces we can follow to achieve our intentions, knowing that the means we give ourselves represent causal traces, that is, traces of the past. Among the two "determinisms" associated with them, let us now ask ourselves which of them is the most realistic!

Whereas increasing entropy brings about dispersion and unpredictability, the Law of Converging Parts talks in terms of bringing order, and gathering like to like! Which assembly would you prefer to elect for deciding how the universe should behave? The one that preaches disorder and dispersion, or the one that does everything to re-establish the proper order of things?

Even the end of a life, which in the reverse direction of time corresponds to a birth, has nothing to do with the drama of death. It is merely the "shrinking" of an individual accompanied by a genetic reorganization of all the information characterising him, in other words, a transformation where nothing is lost. Surely this would be the best way to die?

The difference between a law which consists in increasing disorder, without proposing any provisional calculations other than statistical ones limited in time, and a law which consists in increasing order, even though we do not yet know how to calculate its rules, is similar to the difference we would make between two friends. The first one asks you to follow him, declaring that he does not know where he is going, because his route depends entirely on chance, while the second declares he is going to the same place as you are, because the more the merrier!

Who would you trust the most?

Léo Rosten wrote: "If we don't know where a path is leading us, then we can be quite sure it will take us there."

The resulting question "But where?" is moot here, because the answer is "Wherever we want it to take us", knowing that this place must first emerge from our intentions, in which case it is better to be accompanied by the second partner, with whom we always agree, than by the first. Otherwise, it would be tantamount to delegating the realization of our intentions to the person who lets himself be carried along by chance.

However, if we call causality too much into question, will it not risk affecting our behaviour just as absurdly as if we acted only by chance?

Indeed it will, for inverted determinism does not call causality into question at all, it only completes it. Taking into account time's reversibility, traditional determinism, which operates in the direction of time and is the same as causality, is still a partially valid concept. It is only when it is in itself excessive that it leads us to struggle against the belief that time favours disorder and engenders chaos and unpredictability. This, in any case, is the current credo among the majority of scientists, confronted with the absence of any law of information to explain the choices the universe will make in the medium to long term, for these choices are incalculable.

The unpredictability of the future leads them to invoke chance, chance which puts forward the premise that the universe compensates for uncertainty by means of random choices: the choice for a tornado to pass over one place rather than another, the choice for a rock about the moment it breaks away from the cliff face, the choice for a die to land on one side or another, and so on. For them, these choices have nothing to do with any sort of intentionality, but are a matter of pure chance. Some invoke God, others determinism. We are back to the eternal debate between Einstein and Bohr, when Bohr retorted: "Einstein, stop telling God what to do!"

To escape this debate and make these "choices of the universe" dependent on our intentions, we would need to be able to act upon processes which happen by chance. Visibly, this is not the case, except perhaps in a very small way, if we are willing to trust the results of numerous parapsychological experiments (2) (25), particularly psychokinesis, with random number generators. However, in this book I will avoid the hypothesis of any eventual "psychic" influence over matter, for the simple reason that this notion of "influence" ascribes its hypothetical mechanism primarily to the realm of causality. In any case, we have no need to invoke it here.

We are not talking, then, about influencing chance, but even so a faint subtlety does emerge from this idea, like a first rift in time: *if we could make our own choices depend on the result of chance, without actually influencing chance itself, then our life could inherit its rightful indeterminism, and our Tree of Life would automatically become more diverse!*

Now, is this reasonable? Obviously not, for this would amount to acting in a completely incoherent and disordered way. What a ridiculous idea!

However, I must point out that such an attitude, whether it be ridiculous or not, would have the advantage of producing eventual traces of our future, but what traces they would be! Traces of our own madness? Absolutely, traces of an entirely uncontrolled future, because we would have delegated our free will to pure chance!

For example, if, for reasons of economy and excessive confidence in chance, I decided that every time I took time off work I would roll the dice and only go on holiday if they threw up an even number, then the non-causal result of throwing the dice would necessarily be a trace of my future, since it could not be a trace of my past; for the indeterminate result can only be predicted if my future is known. Whereas it would always be unknown.

According to our Theory of Time, such evidently absurd behaviour, relying on pure (indeterminate) chance, can only be determined by the future. It depends, therefore, entirely on the Law of Converging Parts.

Acting according to "clues" dictated by chance, therefore, would indeed amount to following traces of our future, but it would be an unidentifiable future, and we could no longer attach any meaning to any coincidences which might emerge from it!

That being so, it holds no interest for us, and is quite literally nonsensical. Here we can see the real danger lurking for anyone wishing to trust to such traces!

Whether we make use of it or not, chance seems therefore to be implicit in our quest for traces of the future, in more ways than one, but we evidently lack the keys which would help us eventually take control of it.

If we took into account the order induced by the Law of Converging Parts, then might not this order be exactly what could give us such keys? For the moment, we have amassed the following clues:

– creation of a non-causal synchronous order,
– intervention of indeterminate chance,
– intentions which are expressed, without knowing how to realize them.

As yet, though, we do not know how to use these clues because we still have no idea what the mechanism is that brings them into play. This mechanism is the object of causal logic whose existence we have discovered, but which we have not yet mastered.

As it is, however, it reveals a first rift in time: one which allows our free will actually to exist, that is, to modify our already mapped-out future and draw a new one, outside of any causal conditioning and even if, for the moment, we have no control over it.

Chapter 4—Summary

Our first search for traces of the future leads us to characterize them through observation of a non-causal synchronous order, like the one that characterizes a coincidence produced entirely by chance.

A second search for traces of the future, consisting in the elimination of any causal explanation that might confuse them with traces of the past, suggests we use our free will in a certain way that consists of:

– avoiding any project with the object of realizing our intentions,

– making our choices depend entirely on chance, for example by playing dice.

From this, we conclude that, although it more often than not constitutes an absurd form of behaviour, it is likely to generate real traces of the future.

V.

NON-CAUSAL LOGIC

*In which we find to our surprise that in generating favourable chances,
the Law of Converging Parts cares nothing for their infinitely
small probabilities.*

"Oh! Philippe! Why didn't you tell me you wanted to go up Mt Jouere! I'd have gone with you! There are no paths up there, just animal tracks! You're crazy, Luce told me how you almost spent the night out in the open without any gear. Hah! What a rookie! What do you think we're here for? How did you manage?"

"OK, Jo, it's all right, no harm done, you know I always manage to sort myself out, I'm lucky, otherwise I'd never have risked it. After all, nothing's ever happened to me out walking, whereas I've had lots of breakdowns on the motorway, for example. Frankly, I've had enough of that, and if I had to choose, then I'd rather be stuck out in the wild, it would be much more pleasant, don't you think?"

"You're crazy! You've no idea. If you only knew how many risks I took when I was young, especially out hunting! I nearly died, almost fell down gullies, in fact let me tell you, just hang on while I remember... it was fifteen years ago, and my brother-in-law..."

...And Jo started telling me all about his previous lives. Each time, the outcome of his stories hung on a thread, just like my recent tale. In fact, he had lived through many more favourable strokes of chance than me, but he never called it chance. It was always either his intuition, or instinct, or luck, or skill, and when things turned out badly, it was because something had gone wrong, someone had been at fault, as if under normal circumstances, everything should function perfectly, like magic, as long as no one made any mistakes! As if people were at fault, responsible or incompetent as soon as the magic of life stopped working!

Jo and Luce wanted to share their overflowing love for life with everyone so much that in one sense, it was hardly surprising they thought it normal that everything should just work by itself, as if love ought to operate like oil in an engine. But they could not oil all life's moving parts, and that was their great sadness in life...

In fact, it is generally the great sadness in life of everyone who is full of love and goodwill, everyone who happens to reason with the best of logic and behave according to the best moral codes, without it necessarily causing things to function according to the ideal they yearn for!

This is why we must learn how to cast off our systematic tendency to reason in ordinary time, using causal logic that is ineffectual because it only partly determines what happens!

This tendency of ours comes from our finding it so difficult to reason by inverting cause and effect, due to our ignorance of the Law of Converging Parts or the **law of attraction of temporal lines**, as we will see in the final section. We find it hard to apply this law because it is difficult for us to admit that it will always manage to make chance act for the best. And yet it does not itself seem to depend on chance, which, on the contrary, seems to be linked to the usual direction of time, to causality, that is.

To learn how to apply non-causal logic, we will now examine a "hidden" subtlety of time, which resides in the immense reservoir of possibilities thrown up by the vagaries of nature.

There is a good way to get back to a given present from a given future that avoids having to appeal to a hypothetical Law of Convergence for which we possess neither methods nor rules of calculation. It entails calculating all possible futures from this present, as many times as is necessary until we finally obtain the imposed future. Thus we replace the inverted determinism of the law of convergence by multiple attempts to go back to the future. Why multiple attempts? Because our calculation has to take into account the variability engendered by all the possibilities of causal indeterminism, in particular the external agents that are likely to be present in the environment and intervene by chance.

Nevertheless, most solutions risk missing the target and never reaching the imposed future, and it may even happen that no calculation at all will lead us to this future, even one employing a very remote chance. But if we really take into account the reality of indeterminate dispersion, which can generate a host of possible futures in sometimes infinite variations, we should, by using an orgy of calculations, be able to make the correct solution emerge, even if it is extremely improbable.

We can be helped in doing this not only by dispersive vagaries (vagaries due to the dispersion of possible trajectories), but also by the external agents represented by all the multiple possible interventions of chance.

In the category of vagaries, we can take the variability of the trajectories of the elements present, such as how long the roe deer stopped for or how quickly I came down the cleft, things that might, for example, make us pass the same place several minutes sooner or later, leaving the possibility for the timing to be adjusted to the exact second that might prove useful to a synchronous encounter.

In the category of external agents, we can substitute my flight of partridges for a passing jet, the noise of the wind in the leaves or of an animal in a bush, or even the blinding glare of the sun. To make them intervene, the dispersive vagaries of these agents of chance must, of course, be compatible, or apt to show themselves within the right window of time.

The possibility of dipping into an infinite number of scenarios, thanks to a huge set of vagaries likely to occur within windows of time susceptible of producing a non-void intersection, making it possible that there might be a smaller window in which they might synchronize, gives us hope of finding at least one willing to conform to our imposed future.

Even though it consumes enormous amounts of calculations because of all the vagaries involved, we do now have a method for applying the Law of Converging Parts without even knowing its rules or formulae!

Remember that this method may produce no result at all, that is, it might draw the conclusion that there is a real impossibility of realizing a given future. So we can only exploit it if we help it to help us, by choosing an adequate environment. This sounds a little bit like the proverb "The Lord helps those who help themselves."

Now, to get a better understanding of the importance of external agents, note that I did not actually need the flight of the partridges in order to follow the deer. I would only have needed to pay enough attention to my surroundings to find the way out. This means that because of some determinist conditioning inherent to my personality, an external agent was absolutely essential to "wake me up".

We see then that, paradoxically, for chance to work properly, what is apparently the simplest solution—the most causal, the one that does not cause an external agent to intervene—must be rendered improbable. It then follows that the harder an imposed future is to realize, given the dearth of agents present, the more necessary it is to add such agents to increase its probability, from a non-causal point of view, of course.

The twin presence of the roe deer and the partridges thus increases the probability that the Law of Converging Parts will find a solution, whereas on the contrary, from a causal point of view, it is precisely this addition of an extra agent (the partridges) which gives us the impression that the phenomenon is even more improbable!

The origin of this paradox lies in the fact that, on the one hand, we reason within a causal framework that makes us introduce the very faint probability of synchronism between several agents, while, on the other hand, we reason using non-causal logic, whereby the probabilities of such synchronism lose all meaning. They become redundant, because the synchronism is imposed.

The only pieces of information that need to be taken into account in non-causal logic are the overlapping windows of time for the indeterminist dispersion of the vagaries of each different agent. For example, if we consider the windows of time during which the partridges and the deer are respectively likely to fly off or appear briefly on a rise of land, it is the duration of their intersection that counts, not the probability of the two events happening at the same time. The longer this duration lasts, the more probable the convergence of these two vagaries becomes in helping an event to happen.

All we need to retain is that the weaker the probability an imposed future will happen through causal means, the greater the probability that chance will intervene in the presence of agents likely to facilitate it. This inversion of probabilities has the merit of giving us a glimpse of why, in spite of appearances, non-causal logic might well be quite as rational as the other kind.

The subsidiary question is one of knowing which one of all the possible agents nature (that is, the Law of Converging Parts) will choose. We could reasonably surmise that it will choose the most realistic, the one where the intersection of overlapping indeterminist trajectories will be the broadest. But it could also be the first agent to appear to the Law of Convergence, or rather the last one from a temporal point of view. We will leave that hanging for the moment.

Now let us look at an example that we will subsequently enrich by adding in a human factor, to demonstrate the extent of the power of non-causal logic.

Let us take our broken glass again, shattered into a hundred scattered pieces, and suppose that its falling was the vagary leading us to the future event we imposed upon ourselves. First, imagine that our process of reconstructing the glass misses one of the pieces in its calculations, because of a very strange trajectory that takes it outside the search zone:

this is conceivable if, for example, the lost piece suddenly gathered speed just after the glass shattered and then bounced off several obstacles before landing who knows where...

As we move back in time to the moment, the glass was being blown into existence, we find that we do not have enough molten glass, because of the missing piece. Up to this point, we simply register a rather strange loss of information. But things start going seriously wrong when we consider that somewhere, before the glass has been blown, there remains a small piece of broken glass that has not rejoined the original glass, before the manufacture of this glass has even occurred.

Which gives us a piece of glass that has emerged straight out of nothingness! So there is a loss of coherence when we recalculate any past that does not reconstitute precisely the same glass with all its original pieces, even the tiniest ones.

We can see, then, that the Law of Converging Parts is duty-bound to be highly demanding when it comes to inverse determinism, in contrast to the huge possibilities of the vagaries of causality in the direction of time, which allow the pieces to be scattered according to millions of possible combinations, provided one is demanding when it comes to precision, yet without any of the different dispersals being incoherent.

Now imagine that I impose a future on my programme for inverse determinism in which a piece of glass is expelled in the fall and lands in a shoe lying at the foot of my bed: it is quite possible to envisage such a case, since we have billions of possible combinations.

However, because the glass was originally standing on my bedside table, while the shoe was lying on the floor on the other side of my bed about two metres away from the table, there is a very small probability for this scenario, as the very few pieces of glass to actually reach my shoe will generally have too low a trajectory to get into it. What is essential here, though, is that the trajectory remains a possibility.

This scenario, then, is one of the probable futures of a scattered piece of glass, and we can suspect that for such a scenario, the Law of Converging Parts will have to employ a very powerful programme to avoid the piece in the shoe being forgotten, and making it find its way back to the intact glass. Most importantly, the piece will have to bounce off the edge of the bed and absorb sufficient energy to regain enough kinetic energy to get back to the glass.

Imagine further that I add a human factor to my programme and that as well as the pieces of glass, I impose on the Law of Converging Parts the fact that I have a miraculous escape from an air crash, since I had a meeting abroad that meant my taking an early morning flight doomed to crash.

Frankly, it is quite out of the question that I should die like that! What, then, could the Law of Converging Parts do to save my life?

I never set my alarm clock. Exceptionally, that morning, I use it because I have to get to the airport very early. The alarm startles me and, sitting up sharply to switch it off, I knock a glass off my bedside table onto the tiled floor. I take time to clear up briefly, cursing the mishap which bodes ill for the rest of my day, then I take a quick breakfast while pulling on my shoes and clothes. I leave the house and start down the steps to the drive, when suddenly I feel a sharp pain in my foot which causes me to stumble and fall down the steps. I manage to pick myself up and keep limping down the steps but trip again. After experiencing a sharp pain my twisted ankle quickly starts to swell. Driving is impossible! I sit down to weigh up the situation, especially the consequences of missing such an important meeting. Finally, I make my way pitifully back inside and ask for help. I spend the rest of the morning in bed.

Imagine my surprise when I turn on the television and learn that the flight I should have caught has crashed into the sea with no survivors!

It must be noted that many other external agents could have saved my life: a traffic jam, a bout of fever, a phone call from a friend who dreamt about the plane crash, etc. So now I come to a remark which accurately characterizes non-causal logic: the probability of a piece of glass ending up in my shoe may well be tiny, but it does not seem to intervene in the choice of agents. In fact, whatever the agent, in this case the falling glass, the convergence of parts comes about whatever happens, that is, the piece of glass automatically lands in the shoe: the possibility merely has to exist.

Let us also note that, contrary to how the piece of glass landed, the probability of my breaking the glass as I sat up suddenly in bed because of my alarm was very much higher! Could this be the reason for choosing such an agent? It is preferable to doubt it, for agents can come one after the other or be substituted for each other according to a logic which is impenetrable since apparently independent of their causal probabilities.

For the moment, then, the determining factor, the one that defies probability, has not as yet been identified. All we can say is that the more powerful the factor is, the more likely it is to get around infinitely small causal probabilities concerning the agents of chance in order to impose its scenario.

So what is this factor?

To discover it, first we will seek the origin of the trace of the imposed future, and then we will have to identify what makes this origin so powerful in defying probabilities, which will then lead us to the end of this book. But it is worth the effort!

There are some people I know who, had the story about missing the plane that went on to crash happened to them, would have kept the famous piece of glass that had hurt them so badly, and have put it on display so they could proudly exhibit is as the "lucky" object which had saved their life.

It is so easy to project a positive virtue onto this object that surely it would be the best representative of a trace of the future we could possibly imagine? And a material trace at last, into the bargain!

What is the objective reason which would make us tend to qualify this piece of glass as a "trace of the future"? Is it because it allowed us to escape a tragic future?

In fact, it is the coming together of this highly improbable chance and its "apparent intentionality" (saving my life) which makes us think of it as a genuine trace of the future we would like to put on display.

Let us rephrase the clues we have collected up until now, susceptible of characterizing traces of the future:

– non-causal creation of order,

– intervention of indeterminist chance,

– apparent intentionality of chance.

This means that a new and meaningful coordinating element has been added to the synchronous order we identified previously, this being the apparent intentionality of chance. This too defies causal logic, but as we have just seen, causal logic cannot explain this intentional order either. As for real intentions, they do not even make an appearance in this fortunate plane-missing phenomenon, or if they do, then it is entirely unconscious!

To conclude this first part, we can point to **the obvious independence between the non-causal logic of an event and its apparent intentionality**. For at no time have we had to make intention intervene in the process of converging parts that would be able to explain the event.

If intention does indeed wield influence, then it cannot be material, for it does not intervene in the sequence of events; and yet it seems to be related, on the one hand, to the formation of the imposed future and, on the other hand, to the very faint causal probabilities of the event, about which the Law of Converging Parts knows nothing.

Could intention be at the origin of this second causality, if it does not actually intervene in any way in the process of achievement?

Chapter 5 — Summary

The highly improbable and sometimes "magical" nature of certain coincidences stems from our inability to comprehend their mechanism, their inherent logic.

There is a non-causal logic for which this kind of event entirely loses its improbable character, for on the contrary, it is precisely when we are in the presence of multiple potential vagaries that our chances increase of their coinciding in time and/or space to act in our favour, whereas according to causal logic, the reverse is true.

Even though it does act in our favour, non-causal logic seems nonetheless to accomplish the work of the Law of Converging Parts independently of any intentionality. It is merely obeying the orders of a future which already exists.

Out of ignorance of this non-causal logic, we sometimes attribute apparent intentionality to chance itself. Chance, though, has nothing to do with the mechanism of happy coincidence, in which all that matters is the reservoir of possible outcomes, out of which the one that achieves the potentiated future will automatically be selected.

If the power of intention does exist, then it can only be linked to the potentiation of a given future, and not to the mechanism of its achievement.

PART II

DOUBLE CAUSALITY

VI.

TRACES OF THE FUTURE

In which we discover that our intentions can cause effects
in the future which in turn become future causes of present effects.

"I hope you've got it this time!"

With a beaming smile, Danièle showed me the bill from the bar-restaurant where we had just had lunch.

"22 euros! Another 22! Hang on, I'm going to take a photo of it!"

We had just passed a car registered in the 22 department, shortly after having left the D22 road on our way to lunch in the restaurant in La Javie. And the day before, I had been explaining to Danièle, after taking a photo of a milestone on the D22, that I tended to stumble on series of the number 22 at decisive moments of my life. So I was amused, but also quite intrigued by this new number 22 Danièle seemed to be pointing out to me and which supported what I said, for I was wondering just what might happen that day that was so important!

She burst out laughing.

"No, no! That's not it at all! Look at the name at the top! You really must write that book of yours!"

She had not noticed the 22 from the bill! She had just been showing me the name of the bar-restaurant:

"The New Novel."

I was transfixed on seeing the name, for here was yet another literary reference! It would not have been so astonishing had I not experienced a coincidence the previous evening with the same literary significance, just after telling Danièle that I intended to write a book about synchronicity. We were on holiday in the geological reserve of Haute-Provence, and these were just the sort of oddities of life I wanted to experiment with by constantly putting ourselves in unexpected situations. In fact, I was starting to have an inkling about their mechanism and how to provoke them. Not wishing to worry my partner, I had told her the whole story so that she could be a party to my eccentricities.

One of these had been to decide to take her to a restaurant in Digne the previous evening, even though we had everything to hand for a pleasant meal at the hamlet near Draix where we were staying, about fifteen kilometres away. We were tired, and it was not really sensible. Although Danièle was perceptive enough to sense that I had to be left to follow my impulses without her trying to oppose or even understand them, I was going too far this evening, which is why I explained it all to her.

The restless nature that made me seek out the unexpected in order to provoke synchronicities was working perfectly, if only because it had added a real "plus" to our holidays, which had been enriched in this way by a variety of unforeseen and often enchanting situations. The gîte we had found at the end of the D22 in Draix lay in superb surroundings and we had been given an excellent welcome. We really had no need to go into town for dinner that night.

To make things more complicated, there was a street-market taking place in Digne when we got there, and it was very difficult to find anywhere to park. For extra comic effect, the car was loaded up with too much stuff to fit in the boot, and I was worried someone might break in, something that had already happened to me in Marseille. After going round in circles for five minutes trying to find a parking spot visible from the restaurant, I finally found one, probably the only one free in a 500 m radius! This was lucky, because I was starting to regret my decision to dine in town, and I was worried my partner would lose confidence in my deviant initiatives.

At last we were sitting in the restaurant, and as we were having a pre-dinner drink, she pointed out that we were parked just in front of a bookstore.

"Have you seen where you're parked?"

The bookstore instinctively reminded me of my "intention" to write, and Danièle had felt the same thing as me.

"That means you have to write your book!"

Although I was indeed wondering whether I had in fact provoked the phenomenon by awakening this intention, all the more so as it was exactly what I was seeking for my book, I still had my doubts. And I must admit that because of this persistent doubt about denouncing predictions, I could not help looking for some other explanation that evening, by calculating the basic probability of finding myself parked opposite the bookstore. Maybe I had unconsciously been looking for it? Had it really been the last available place?

So it was not until the next day, in the Pizzeria in La Javie, after a whole avalanche of synchronicities I will describe in their entirety in the third part of the book (the only ones I quote here are the two final coincidences), that the note held out by Danièle came as the coup de grâce convincing me not to delay a moment longer in settling down to write this book, as if the "universe" had ordered it. On my return from holiday a few days later, I actually got started on it.

And as if to signal to me that on that day, by taking the decision to write my book, I might be about to change my life, the final coincidence in La Javie was accompanied by a series of number 22s!

I put forward here the hypothesis that the examples I experienced constituted real traces of my future, but this time with an obvious intervention on the part of my intention, probably amplified by a very real willingness to open myself up to its effects.

These sorts of manifestations of chance, which we think we might be able to qualify as genuine "traces of the future" remotely controlled by intention, present a very interesting peculiarity: these traces are potentially controllable!

The manifestation of such traces necessarily implies that some intention must exist, even one full of doubt, if only because without any such intention to help us interpret traces, we would be unable to distinguish them from chance.

If this intention is reliable, then it carries with it a genuine potential for being realized, making it likely to produce effects in the future. Thus far, however, we have nothing new: we are still waiting for the universe to be kind enough to give us the operating instructions!

Let us return to our quest for traces of the future, this time examining them in the light of their underlying finality: what can we observe around us to show the creation of a non-causal order with some apparent finality, or with the aim of realizing some intention?

If we put it like this, then the question makes us think about the emergence of certain particularly creative processes in the living world, such as mutations, but also man-made creations, of course, or those of certain evolved living systems, such as how ants and bees organize themselves. Could these order-creating processes be getting some non-causal support?

Without life, the universe would suffer a systematic degradation of its state and components, due to the overall increase in entropy over the course of time. Effectively, then, we may wonder if there might not be non-causal processes to be found among living systems, which show an impressive ability to maintain their entropy by absorbing energy from their surroundings (food). For how do they manage to cause such a complex and effective system to emerge?

As far as the processes of spontaneous living creation are concerned, with their apparent reliance on chance, we can point to mutations, particularly those that allow a living thing (plant or animal) to adapt to drastic new living conditions, where lack of food and predators are the two main eliminating factors: the chameleon, or better still, the Kallima butterfly from Ceylon, which both disappear by taking on the colour and texture of the plants they are lying on, or the Ophrys orchid, which so closely resembles the Goryte wasp that the male insect pollinates the flower while mistaking it for its mistress (34). They all survive using stratagems that lead us to believe that the mutations that engender them originated in an intention!

Could the sudden appearance of certain mutations be a trace of the future? Would it be sacrilege to dare mention such intention in nature when Darwin clearly forbade us from committing such an offence?

Examples abound, not just in biology but also in archaeology and palaeontology, that lead us to suspect an error in Darwin (37) and his theory of natural selection. This idea, which has become a age-old paradigm, just like determinism, using the principle of competitivity to influence the entire economic organization of our society, has for origin causality, always causality and only causality, making us elevate the principles of competition and the survival of the fittest to the level of an unshakeable theory, or else commit sacrilege against reason and dogma. Only mutant species which resist their environment survive, which is apparently what explains such prowesses of nature, supposedly with time as its ally, since nature has millions of years to experiment before imposing its final selection.

Do things always happen like this?

Even today, we only understand mutations within the context of this theory, which relies on the exaggerated intervention of chance, and this is precisely one of the reasons the theory is becoming more and more controversial.

Excessive intervention of chance permeates statistical physics, then, just as much as it does biology. Might this not be a sign that scientists sense how important chance is to science, but that they have not put it in its proper place?

To put chance back in its proper place, let us now consider man-made creations, still on our quest for traces of the future. We could point to all man's technical constructions, all his artistic or architectural achievements. Simply take one of the diverse material elements of one of man's creations, such as a brick, a constituent part of a house, for example. Is this brick a trace of the future? It would seem so, considering that the brick will be integrated into a construction in the future. But what about the unused bricks? Do elements exist that, considered in isolation, cannot by nature maintain this state of separation and must thus, of necessity, form an integral part of some future creation?

We understand that by themselves, bricks and other component parts of a house are not enough. The intention to build obviously has to exist too. Just as there was an urgent need to adapt in the case of mutations, human constructions require the presence of an individual with strong motivation. Put real human need in the presence of all the material elements likely to satisfy that need, and the construction will happen.

But does this really mean that we will be looking at a set of traces of the future?

Well, no, it does not, for we have to reject needs that are filled in a causal manner, that is through the intervention of a project already planned in advance. The brick only bears witness to a past where the need to use it already existed, before it was assembled in building the project. The builder picked it up in an entirely causal way.

In order for this not to be the case, the project for building a house would have to emerge spontaneously, independently of the past, as if, happening upon the brick one day, our builder then devised the plan for building the house all by himself!

The realization of the project would then seem to emerge like a mutation, that is by chance. But is this really the way we realize our projects? Surely our plans are quite simply already there, merely forgotten, and then awakened by a trace?

Even if this were the case, and whether it be a question of human construction or mutation, is not chance always there alongside an awakened intention, reactivated in the presence of a trace of the future?

Take, for example, the chance discovery of a treasure that allows its finder to realize a project he has dreamed of but had not the least apparent intention of accomplishing, not possessing the means to do so. In this case, we may consider that for this person, chancing upon such a treasure constitutes a trace of a future where the thing the person most desires is going to be realized, the trace being not the treasure itself but the discovery of it. On condition, of course, that the event not be artificially provoked by someone generously wishing to put the treasure in the hands of our discoverer!

In this type of case, where an intention appears to be "awakened", as opposed to apparently unintentional traces of the future, or ones whose intention is hidden or unconscious, we see that once again, it is always chance that makes information somehow emerge about a new potential that did not exist prior to the discovery. Indeed, had our discoverer earned the money he needed by working, or if he had received it from someone else, then the information "individual possessing a sum of money allowing him to realize his dream" would merely have been a trace of the past, and we would have invoked causality to explain the fact. Chance, then, allows us not to place in the category of "trace of the past" a fact which is crucial in modifying the future.

I refer here to "modification of the future" because we are quite happy to admit the underlying existence of a sort of ordinary version of our life, one we would really have lived had not chance intervened to take us onto another branch of life.

Could it be that some events capable of turning someone's life upside-down constitute genuine traces of the future, once they rely on chance, truly indeterminist chance, accompanied by the now obvious role of awakened intention?

The answer is yes, insofar as such chance eliminates all causal influence. But note carefully that we still have not supposed the least non-causal influence of this intention! And yet, in this sort of event we have already found traces of the future where such influence plays an obvious role! Intriguing, isn't it?

To delve further into the question of the role of intention, let us now consider a trace of the future of someone seeing his dreams become reality, a trace where this time, his intention plays a supposedly active role, from the point of view of the Law of Converging Parts: let us suppose that this imaginary future is "activated" in the sense that its probability of happening has been raised to the necessary level, without our knowing exactly why, at least for now.

To make this imaginary future correspond to the present prior to the discovery of the trace that triggered everything, we find ourselves obliged to play out, in the opposite direction of time, a scenario which is adequate, but which at first glance competes with any other purely causal scenario of failure, or even of success, where the individual's own qualities, efforts or any other assistance he may ask for, are made to intervene.

To force an unpredictable, successful scenario, would the Law of Converging Parts be capable of producing a modification of life trajectories in the universe in such a way as to place our man and the sum of money necessary for achieving his dream in the same place?

What freedom might the Law of Convergence possess to be able to act in this way?

The prevailing answer, as in the case of the broken glass, is as follows: the Law possesses a host of possible vagaries, ranging from chance encounters to the simple natural dispersion of life trajectories, inherent to the indeterminism which presides over the normal process of time.

If the treasure is buried, the Law may even draw our man's attention to the ground with a falling fruit or leaf, if the indeterminist window of time for their falling has a chance of coinciding with our man's trajectory. This means he would simply have passed at the right time and in the right place, and would have suspected that something lay buried just there, where his gaze had fallen.

So it is merely a question of synchronism, providing our discoverer's environment is conducive. If the treasure had been in that environment for a long time, it would have been useless for the Law of Convergence to proceed by any other more complex scenario. The simplest route would have been to change nothing in the structure of the universe and make use of a few vagaries in our man's trajectory.

Who among us can always pretend to know where he or she is going and always walk at exactly the same pace?

For anyone who maintains that there is only one possible trajectory, and who continues to ignore the indeterminism of vagaries by bringing up the objection of the Theory of Hidden Variables, I would point out another factor: in quantum mechanics, until a trajectory is observed and interacts in a decisive way with the environment, it is likely to exist according to a multitude of versions. But we will see more of this further on.

The simple properties of natural dispersion of the potential trajectories of human beings are already generally sufficient for allowing such coincidences to occur, providing that the possibility of them occurring exists in the environment. Luckily, finding oneself in the right place at the right time is generally favoured by many other conditions likely to generate indeterminism or lack of precision. If I allow myself to be distracted by the appearance of a bird in my doorway, I will be a few seconds late leaving the house. If I give a lift to someone whose car has just broken down, I may even be taken out of my way. This, in a general way, is how little things are susceptible of modifying the position in space and time of anybody, at any time. The Law of Convergence thus has the possibility of playing with any considerable number of dispersive factors in order to achieve the programme which has been favoured in the future.

Now let us take an example that will help us discern the potential role of intention in how traces of the future are formed. In order to complete a project which is important to me, I need a key piece of information, although I am unaware of the fact. In fact, it is precisely because I do not know that I need this information, and that I think I already have everything I need for carrying out my plan, apart perhaps from a few details, that I am firmly confident I will be able to complete my project. And this confidence is what lends me the self-assurance that I will know how to manage all the details I need to sort out in order to reach my objective.

It is at this point, quite by chance, that I make an encounter which brings me that vital key element, without which I would never have been able to complete my project successfully: a piece of advice resulting from human contact.

How did this encounter come about? How did I thus manage to avoid having to learn something the long way round? Did I, unawares, raise my chances of success by doing so?

Evidently, most of the time we are under the impression that we cannot programme our future, especially in those particular cases where we do not have the means to realize what we wish for.

However, if we consider our Tree of Life, the route which leads us to this realization already exists, even if it is very improbable. The problem is that it coexists along with other routes that lead to failure and are much more probable, so long as we do not hold the key element for success.

The solution, then, would be to increase the probability of the route to success until it became more probable than the other routes.

Let us just take a look at how this sort of probability could simply increase.

Let us suppose that now we are aware we are missing one key element, but that we keep our intention to realize it steady, because we are determined and because a little voice inside us is whispering,

"You can do it!"

Such an intention, reinforced by our own belief, would have the advantage of eliminating any scenario of giving up, even before the factors triggering them showed themselves, which would result in the probability of a successful scenario being immediately increased: so does this mean faith could be an active ingredient?

Let us now suppose that, armed with confidence, we decided to make deliberate encounters because, with a little luck, we might chance upon the one capable of bringing us the key element. So just like that, even before we started making these encounters, the probability of achieving our objective would be increased further, for we would be giving ourselves additional openings: we would be increasing our chances.

Now let us imagine that the Law of Converging Parts be "incarnated" as it were, not by unknown equations but by some kind of an organizing entity. He would study the existing elements, and notably any possibilities for playing on the indeterminism of the trajectories of people in possession of the key information I need, and then he would choose the trajectory which led to our meeting; including among those trajectories, if necessary, missed alarms, breakdowns, incidents of all kinds, etc, anything that might delay or deviate the trajectory of one or the other of us, and all depending on practically nothing!

The motto of this kind man, this executor of the Law of Convergence's charitable works, would thus be: "If it's possible, I'll do it."

Of course, objections will be raised against this mechanism of chance, accusing it of not being based on anything tangible, and accusing me of having up merely enticing the reader to this point with a hypothesis no one would wish to reject if it could possibly be true.

This is not quite accurate, for we would be forgetting what I have just explained, which is that **our intentions constitute an instant amplifying factor of the probability of chances favourable to their realization**.

What does this factor consist of?

On the one hand, it consists of our free will and its ability to make a reliable intention emerge from among all our various intentions; on the other hand, it consists of the ability of our intentions instantly to modify the distribution of probabilities on our Tree of Life, providing, of course, that they are powerful enough to induce faith and confidence.

What we must now concentrate on is the fact that this modification will also have the effect of instantly modifying the already-existing structure of future space-time by means of the emergence of a new potential on our Tree of Life.

Which means that **our intentions are active not only within our own brains, but also somewhere in the universe outside our brains!**

To clarify this possibility, we will propose a new metaphorical representation in the following pages. We have agreed that our "visualizations of the future" happen in the present, but they seem to have an instant incidence on the probabilities of realization of one or other different future branches of our Tree of Life. These intentions should therefore be considered as information that occupies our future while bringing an increased potential to certain zones of our Tree of Life.

In order to illustrate this potential, we will now use the analogy of how water trickles down from the mountains and into streams, flowing into rivers, tumbling down dams or waterfalls, etc.

Since water's potential for falling is indissociable from the ability of its current to carry information, we will therefore use the image of the current to illustrate how the information we need to survive—that is, the information our free will needs in order to control our intentions—is transported.

In this way, water will be the vehicle of any information in whose presence we will have choices to make.

Chapter 6—Summary

The mechanism by which our intentions are active and therefore susceptible of generating traces of the future is as follows:

– when our intentions are sustained by faith or confidence, they instantly increase the probabilities of realizing a future already potentiated on our Tree of Life;

– if a causal route does not exist, or no longer exists, then the potential of non-causal solutions for the realization of this future will see their probability increased, according to a route that descends from the future to the present;

– this will favour the emergence in the present of favourable chances, and as a consequence, we must expect to observe around us opportunities for routes leading to this realization, providing we are open to them;

– our intentions are active not only in our brains, but also in a future space-time outside our brains, one that houses all our potential futures: because our intentions instantly modify the probabilities in this future.

VII.

THE CURRENT OF LIFE

*In which we introduce a third model, a network of rivers, to represent
the conditions of our existence and in particular,
all the efforts that oblige us to keep rowing.*

A busy Friday morning. The entire day organized down to the minute into half-a-dozen meetings. The first one at 9 o'clock. Then check to see that the prototype is working for a demonstration in the afternoon, then a meeting at 11 o'clock. Try to find the time to go to the post office and collect a recorded delivery. Lunch with a chargé d'affaires at twelve thirty, then a meeting at 2 o'clock. Wangle it to finish the meeting before the rush hour on the motorway as the school holidays start. Foresee not being able to avoid either the morning or evening traffic jams. Try to appreciate the time I waste waiting in them, because it is a source of contrast that enhances my luck. At about 4 o'clock, all being well, I can escape from Marseilles and drive up to Haute-Provence, past the threshold of the Gate of Time and into a different space-time, a universe where time slows considerably and the silence bears majestic witness to the fact. This Gate of Time is the "Pierre Ecrite" mountain pass, the entrance to the geological reserve.

Three pm. The person I am talking to is almost astonished to hear me reassure him while at the same time approving all his comments. "No problem." "Please don't worry about it." "You know, we won't know how to react to unexpected twists until we're actually up against them." "If we try too hard to anticipate the unexpected, we'll only provoke it." "Believe me, I know Murphy's Law*, I lived with it for 22 years." At three thirty, I'm off. At four-thirty I finally manage to get out of the crush. Suddenly, a female voice talks to me: "On-board control centre. Speed limit 110 km/h." I'd forgotten about that one. Nothing to report, no one was there, and I had already reduced my speed in any case. The mere idea that I would soon be changing systems had made me slow down. If they had stopped me, I would even have taken the time to chat. By the time I get to the Pont de Mirabeau, we are already in a different world, even if we are still on the

83

motorway. Everything becomes beautiful. And it is important to give oneself time to arrive at one's destination, especially when that destination changes one's notion of time.

I drive through the mountain pass. At the first hamlet, a hand waves to me from a distant house and I wave back, guessing it's Gino. On the road, I pass Bernard's tractor coming the other way. "Hello Philippe, you haven't seen Maguy by any chance, have you?" "No, no, she might have called in at home but I've only just landed, as you can see." I say it as if I were coming back from a trip to the US or even the Moon. Bernard senses that I am still a little out of sync, and tells me I have edged my car into the ditch. We say goodbye. Just after the village, I have to stop again and help Nicolas bring his herd of cows back from over the road where they have been grazing on the lushest grass. A few minutes after this small spurt of action, I pass Annie, the shepherdess, in her 4L truck. We kill our engines and have a chat for a good quarter of an hour. We are blocking the road, but it doesn't matter. It's a question of habit. We tell each other about our week, chat about today's weather and what the forecast is. The weekend is already planned. Tomorrow, drinks before dinner. The day after tomorrow, a stroll along the Monges Hills. Ah, there's a car behind us. Oh, it's Gilbert. Without any hint of impatience, he waits while we finish our conversation. We give each other a wave and he signals to remind me I have to go and see him. We take the time to admire the situation. It's rare to see three cars at once blocking the road. It's worth stopping and taking it in properly. This will have added up to a half-hour hold-up in all on the Road of Time. Making it all the more pleasurable. When I reach home, I collect my post from beneath the bird box, and spot Jo rifle in hand, coming back from hunting. There's quite a crowd out this evening. Tomorrow evening, I'm invited to a game supper. This evening I'll simply bring in the wood. My weekend has just been organized completely by chance. Whereas a few hours earlier my day was starting out following a timetable which felt like just so much stress! What a contrast! But I knew it. I knew that when I passed the Gate of Time, I was leaving the current of a well-filled life, the current that forced me to go through compulsory, daily, systematic stages that were limiting and often planned long in advance. In that current, chance was banned, only just about tolerated for the sake of a flat tyre, for example, but not for one's alarm failing to go off.

As I passed the Gate of Time, I knew that chance was becoming my ally!

During the course of such a day, we observe two types of meetings, causal or non-causal in nature, where the encounters in the second category are apparently produced by chance.

So we can classify these meetings into two types of "information" about our future, allowing us to control our lives and take our decisions:

– information of the causal type, traces of the past allowing us to take these decisions according to our skills and desires, but also according to our conditioning or our life "programmes";

– information of the non-causal type, traces of the future contained in the chance of opportunities, coincidences, intuition, susceptible of lending a helping hand in the exercise of our free will. We suspect that such traces depend on our intentions, but this is still just a hypothesis.

Here I would like to point out that our ignorance or disdain for the second sort of information leads us to make exaggerated use of the first sort, resulting in our being trapped within an excessive current: the current of life.

Could all this information really be transported by a current? This is another question we will attempt to answer here.

The "current of life" is an almost poetic expression which qualifies a rather negative and passive aspect of life, describing the fact that every day we are swept along by a set of necessities, by all sorts of work and obligations that may give us the impression of being "carried" towards a destination against our will.

In a more neutral way, we also talk about "time flowing", which also suggests a current that accompanies us in the direction of time. Time itself is said to produce this current of life, carrying us along towards an uncertain future. We often encounter the metaphor of the "course of time" in literature, too, implying that time takes an irreversible course which supposes that it flows from the past towards the future.

However, according to inverted determinism, this current related to the flow of time is instead orientated towards the past, suggesting a reverse current!

In contradiction to our theory, might there exist some physical property to confirm the notion of time running towards the future, which would thus raise us automatically towards the summit of our Tree of Life, or even up onto the heights of the Road of Time?

Or are there, on the contrary, in accordance with our theory, any objective reasons to think of a flowing in the reverse sense of time?

Let us consider how the fluids in our models behave, starting with how water, dew or rain flows along the branches of our Tree of Life: water flows downwards because of gravity. Perhaps this flowing towards the bottom, towards the past, according to inverted determinism, might contain something interesting: that perhaps it brings us back information about our future which, transported by the water, would spring up in the present in the form of traces of the future?

Let us suppose that time behaves like a flow that, on the contrary, transports information towards the future, at the same time as taking us there ourselves. So how is it that when we just let ourselves be carried along like a bottle on the stream, we invariably follow a river, then a larger river, and eventually end up in the ocean? Could the sea really be our unique, final and entirely predictable destination?

In fact, doesn't the sea symbolize our birth instead? Surely the innumerable springs that produce the water flowing into our rivers symbolize our destinations, like the flowers or fruit on our Tree of Life?

For if we were to follow a river back upstream, from the sea right up to one of its springs, we would find ourselves in exactly the same situation as when we climb our tree of life from our birth, from the trunk up to the fruit: the spring plays the same role as the fruit, symbolizing one of our destinies.

It would seem, then, that our models lead us not only to reverse the direction of determinism, but also the direction of the course of time.

Now such a course of time, running in the opposite direction to our own evolution, in accordance with reverse determinism, would be likely to bring us information back from the future.

Indeed, when we consider the causal and non-causal information we have catalogued previously, we notice that a flow against the course of time would correspond perfectly with our going towards all our meetings, whether they are planned or unexpected, for they take place at fixed moments and, in contrast to ourselves, stay frozen at a precise date. So they pass from a future to a past state, moving past us in the opposite direction, for we remain in the present. All our meetings, all our opportunities, all our encounters are fleeting moments which pass us in time exactly like boats carried along on the current, while we ourselves are heading upstream, against the current.

This productive correction of our metaphor of time flowing against a current does, however, bring up what appears to be a problem: it apparently brings to bear upon us a determinism from the future which is just as restrictive as the one from the past, if not more so. This would mean that we were "stuck" in both senses of time. For whenever we were faced with a choice undetermined by our past, it would be determined by our future!

How can we avoid being trapped in this way between our future, on the one hand, which by already existing brings about our present in a determinist manner; and the consequences of our past on the other?

This really is a problem, isn't it? Don't we sometimes have the feeling that not only are we conditioned by our past, but also obliged to turn up for all the future meetings heading towards us?

So we see that our destiny weighs on our present in just the same way as on our past, thus appearing to make our free will illusory. Doesn't this make us prisoners in a vicious circle of causal consequences of our past and non-causal consequences of our future, each playing with the others and preventing us from altering our destiny?

The flow of time is thus shown to be a factor likely to thwart our destiny: we have to go against the flow rather than swim with it! We have to resist it, in fact, to stop ourselves being carried away!

How can we break free from these constraints?

If we use imagery to describe our life, then it is born in the "sea mother", and consists of choosing to swim up a river, then a tributary, then a torrent, then a rill, until eventually we reach a source. So in order to have any "sense", our life would be about gaining altitude, whether it's a question of swimming upstream, travelling the Road of Time or climbing a tree. Our three models are working in unison, telling us to do the same thing: make the effort to climb! Thanks to their richness and complexity, this network of sources and rivers, as well as the roads and paths on the Road of Time, offer us a vast potential choice in life.

According to all these models, however, it would seem possible to count the truly important bifurcations in life on the fingers of one hand. A tree, a network of rivers or roads and paths each comprise a relatively limited number of bifurcations. Could these be the principal choices we must make in life? Could these, quite incidentally, be the choices common to us all, the choices in our personal and professional lives?

We do have other alternatives: nothing prevents us from exploiting all the richness of the territory of the Road of Time to avoid making the same choices as the rank and file, the ones that consist of keeping to the roads or the riverbanks.

Everything happens as if keeping to the current of a river or a stream conditions us, by making our important choices, such as a job or a spouse for example, reliant on causal logic. If we do not step outside this logic, then we just keep moving with the stream until we encounter the same apogee as anyone else.

But what is the point of staying midstream in these currents, which are overloaded with information and indiscriminately ferrying constraints?

Does the vocation of the rank and file really have to be to keep on rowing?

This would be to ignore that swimming upstream is not necessarily the best way to reach a source! Clearly, it is the most obvious and reliable way not to get lost! But what about when the current is too strong for us? And once we have ventured fairly far, how do we get up waterfalls? Isn't it dangerous to take those sorts of risks?

Do we really need to follow all this water? After all, isn't it just acting as transport for temporal information?

Do we really need such an over-abundance of information?

Might we be forgetting our ultimate objective—our quest for Theopolis?

Let us keep in mind that if we never depart from a current of temporal information, meetings and encounters, then we will be caught up in a flux of information where we will find it hard to discern the traces that actually concern us.

In the real world, we swim upstream every day of our life against a current of necessary encounters, an incessant flux of people with whom we must have exchanges. And so we spend our whole time "rowing" to obtain our means of subsistence and improve our comfort and safety. These difficulties translate into a quest which has nothing to do with information which could help us find Theopolis: on the contrary, it consists of ceaseless resistance to constraints exacted from us by society in return for earning our means of subsistence or comfort, so that we can trace our path with more security or certitude. This information is money, administrative papers, diplomas, title deeds, brands, copyright, dispensations, passes, etc., all of them adding up to just so much assorted paperwork that we are endlessly having to show to people we meet in order to win the right to advance along the road of our own time.

Difficult, then, with such a huge flux, to distinguish the other information, the second sort, the sort that might bring us non-causal nuggets, the ones meant for us!

Just as we recommend avoiding main roads in order to avoid seeing our life blend into that of the rank and file, here again we recommend avoiding main waterways, so as not to swamp our life in a flood of too many exchanges:

the flood of the current of life, that prevents us from distinguishing the part of the universe that is nonetheless waiting to reveal itself to us.

Whether we are on main roads or rivers, as long as our intentions keep us there we will be led to follow a single path where every exchange we might have will only give us the illusion of free will, trapped inside the first causality and unable to detect the second. Under such conditions, we should not be surprised, therefore, if we are sometimes brought to question the meaning of life.

In our modern societies, we have replaced our fundamental need to enrich and explore our Tree of Life with an inculcated need to progress as far and as fast as possible.

We have quite literally forgotten the vertical dimension!

Our technological progress, itself trapped inside simple causality, has amplified the phenomenon, reducing our tree of possibilities to a few traces that everybody can follow, and obviously making any quest for Theopolis quite redundant.

Evidently, this is not how we will detect any traces of the future that might prompt us to resume our quest for Theopolis if we have not actually abandoned it already.

Wherever there are no longer roads to direct us, nor rivers to impose information on us that conditions our choice, this is where we may find information of the second type, and the rarer it is, the more clearly it will signpost our route.

Otherwise, if we keep to the places where traces of our future are diluted, how can we manage to detect them?

One question still needs to be answered: how can we distinguish the first, causal type of information from the second type, from traces of our future?

Let us keep in mind the significance of water, which gives rise to all these currents: water is the vehicle for information that calls on us to make choices.

Chapter 7—Summary

What we have learned to perceive as the flow of time towards the future must be reversed. For we head towards all our encounters, whether planned or the result of chance, and everything happens as if we were progressing through our life along a counter-current, moving towards the past.

This current is constantly bringing us new information about our future, information both causal and non-causal, and we tend to respond indifferently to it all by making conditioned choices.

The more numerous our conditioned choices, the harder it is to detect the "traces of our future" that might be facing us, because they result in our losing our free will, which prevents us from forming traces as much as it does from detecting them.

So if we want to detect traces of our future, we should flee zones and habits that assail us with information we react to in a conditioned manner (and we can make a start by taking a holiday, for example).

VIII.

"LETTING IT WORK"

In which we discover a special sort of "trace of the future" that we can embrace by "letting it work", a technique that can help us achieve extraordinary prowess if we just let the current "act".

"Just roll your boule, Irène, don't try to play well whatever you do. Your boule knows where it has to go, it doesn't need you to help it!"

"What's Philippe going on about! Aim a little, he's trying to bewitch his partner!"

Irène had never played pétanque before. We had formed two teams of two players by drawing names out of a hat, and two experts were playing against Irène and me. Her boule came to rest ten centimetres from the jack!

"Wow! Well played Irène! Lucky you didn't listen to Philippe, huh?"

"Oh but I did, I did exactly what he said, it's just a stroke of luck!"

Then Jo went straight ahead and knocked Irène's boule out of play! Back to square one! I gave my partner fresh encouragement:

"Go on, play your second boule just up against Jo's. And don't forget, it's the boule that's playing, not you!"

So Irène did, and unfortunately threw her boule too hard and little too far to the left. But by some incredible stroke of luck, a bump in the ground made her boule veer off to the right, knocking into Jo's boule and carrying the jack off with it, towards the spot where her first boule lay. The jack came to a stop next to it, five centimetres away, just before her second boule came to rest right on top of it!

With two superb points in one go, it was now quite impossible for our adversaries to get the point back using a knockout shot!

"I don't believe it! Incredible! You're a magician, Irène! That was an impossible shot!"

But actually it was not impossible at all, because when she threw her boule, Irène had potentialized a multitude of trajectories generated by the indeterminist vagaries of the ground's uneven surface: and the shot became possible. Highly improbable, but possible nonetheless.

Sometimes, in contrast to all the opposing currents that force us to face up to effort, difficulty and confusion, we experience situations in which everything "flows naturally", moments of grace in which, like Irène, we achieve great prowess, moments where opportunities are given to us, without our having to make any effort at all.

At times like that we thank "luck", not believing that we might be responsible for what has happened.

However, we have seen that among all the currents of our life, some might be carrying information of the second kind, or "traces of the future". The question, then, is this: how can we detect the currents likely to have come from a source somewhere in the future, fed by our intentions?

Let us point out that, if we head towards our future and never alter our intentions, then our path along our own tree of life will follow the most probable potential, the one we call our destiny. This destiny will remain determinist and automatic as long as our free will, thus rendered useless, does not modify our actual intentions.

In such a case, where our destiny is all mapped out, can we hope to find traces of our future?

Now we are finally getting there.

A destiny made determinist through failure to exert one's free will cannot tilt towards an alternative destiny unless indeterminist chance intervenes, in which case it can only be a chance independent of our will. Now as far as I can tell, our lives are all interwoven, and what does not depend on us may depend on someone else, or even on the accumulation of all human intention: under such conditions, it seems naïve to hope to "decode" the least trace of our future.

Moreover, we can only hope to find traces of our future if we are ourselves responsible for **some change in this future** that will increase the probabilities of an alternative destiny.

Just suppose that digging down deep inside ourselves, we manage to extract intentions that are at once new and authentic. We have added the condition of "**novelty**" to our quest for traces of the future.

If our intentions really are new, then they will instantly trigger a modification in the distribution of probabilities on our Tree of Life, and these new probabilities will immediately translate into a redistribution of potentials along every path of the Law of Converging Parts.

This will translate into our having created **new sources of information** in our future.

We are now going to ascertain the true power behind the image of a source.

By altering our authentic intentions, it is almost as if we are making it "rain" somewhere in our future, a place where water will then accumulate.

Let us return to the Road of Time and examine the sources responsible for all the water flowing in its waterways.

There is snowmelt, permanent springs from the groundwater, and lastly water from transient springs.

Here we will consider that the first two sources are permanent, for their flow is unceasing, even if its volume varies. Their permanent flow allows us to consider them as "causal" sources, these causes being the snow and the groundwater. There is no more room for chance in this sort of flow, which in some way results from the accumulation of all human intentions!

The transient sources due to rain and storms are, on the contrary, unpredictable, indeterminist and localized.

The fact that they are localized and transient makes them apt to represent the effect of an intention that is localized and transient until it encounters the means to be realized.

The fact that they are unpredictable and indeterminist, like the weather, makes them apt to serve as model for a non-causal process determined by the future, since it cannot be determined by the present.

Let us get back to our quest for traces of the future on the Road of Time.

We can now ask ourselves whether, having formulated an authentic and original intention, the act of coming across a small welling stream would not be likely to supply us with an answer to this intention, or give some direction to our wanderings.

The stream would be like an invitation to follow a new course of our existence, triggered and handed to us by a modification on our Tree of Life.

However, there is nothing really to incite us to follow this course of water, because it might originate from some transient source independent of our own intentions.

We are obviously lacking some information.

Could that information reach us on the current of this unexpected source itself, in the shape of clues to prove that we are indeed concerned by its invitation to follow it upstream, like a guiding principle, "a matter of course"?

As the source is situated upstream, on higher ground, then it must be supplied far enough ahead in the future before the water can bring us any clues. It requires a certain amount of time before the accumulated water is sufficient to produce the information that interests us, coming back in time.

To hasten the process, the impatient among us might decide to dig the waterway themselves, rather than waiting for it to appear by itself. This would amount to taking calculated action, counting on the existence of the source upstream fed by their intentions, and waiting for it to appear, through a combination of chance and a little effort. But to dig that sort of channel would risk falling back into a causal context, as if in the end, all they were doing were paddling upstream. Why not dig the entire channel, in that case! And they would still need to have information about exactly where their intentions had rained down; and taking into account all the uncertainties related to the unevenness of the ground, it would be as awkward as trying to anticipate the successive rebounds that would have to be imposed on the trajectory of a billiard ball to get it back in the pocket. And they would spend all their time missing their target.

This is not how champion billiards players or archers operate!

Do you think that champions at this sort of game, where an infinitely precise aim is sometimes necessary, really have to strive to reach the summits of their art, once they are properly trained?

The whole idea that to attain our objective we have to learn to calculate our actions properly, according to adequate levels of experience and knowledge, is in fact an illusion produced by our education: even after excellent training, most of the time the process of reasoned thought imposed upon us by such an education actually deprives us of the immense skill of knowing how to hit a perfect target every time.

In reality, excellent training merely opens a progressively more restricted field of possibilities to us around the target, but it does not give us direct access to the essential thing: its centre.

In his book, *Zen in the Art of Archery* (8), Professor Herrigel gives a good description of the importance of cultivating an attitude apparently contrary to reason, that of "letting it work", which allows this immense skill to operate: having spent long months learning how simply to draw his bow without straining and to remain relaxed, his progress was still insufficient. He had to wait for years before he learned how to release the arrow. When he finally analysed his success, he declared that he had acquired an ability to let go which gave him the impression that he was "letting something work", something that "fired for him", according to his own expression. He likened this stage of development to a state of awareness that corresponds to the nature of Buddha, which is where the title of his book comes from.

According to him, the role of a master archer is limited to the act of preparation, placing the body in position after ridding it of thought, and "letting It shoot", as if some other entity were taking over, starting from the perfect bull's-eye and coming back to the "perfect shot" at the moment of releasing the arrow.

Considering our observations of rivers again, wherein information can spontaneously arise in the future and come back in time to burst into the present, the "perfect shot" would consist of visualizing the shot at the centre of the target in order to provoke this spontaneous apparition, this upstream activation of a "source of current" born of the future. Thus we put forward the hypothesis that "visualizing" an objective could create pressure upstream, in the future, before our present derives from it.

This would mean that the act of releasing the shot would be triggered by a current arriving from a future in which the event already existed. So everything would occur as if the bull's-eye had been hit even before the shot was released, because its route would be traced, starting at the centre of the target and moving in the opposite direction to the arrow. In fact, according to the Law of Converging Parts, by going back in time we can predict that the arrow will invariably return to its nock on the archer's bowstring—since there is only one coherent solution—whereas a shot calculated in the normal sense of time would have every chance of missing the centre of the target, taking into account the infinite number of unpredictable solutions.

Exerting our free will, then, depends partly on our ability to let go, allowing us to detect new channels emerging from the future, after having ourselves contributed to their emergence by means of a mechanism we gradually update: that of second causality.

If it is possible, then this exertion would allow us to benefit from the vagaries of life and the chaos which accompanies them, instead of enduring them by systematically trying to counter them through perpetual readjustments. It would also allow us to increase our possible choices when facing new paths, but what are these choices? What is it, within us, that makes a choice that consists of just letting things work?

To sum up: letting things work lies, first of all, in informing our future by means of visualizing our intentions, so that we create a sort of pressure upstream. Then we need to allow the current induced by this visualization to take action, which does not imply an immediate result, since we have to wait for the information to trickle back down. Once we perceive this information more or less consciously, then we will be as it were "attracted" by a future we have created ourselves, one susceptible of replacing our former destiny.

Nevertheless, one question remains in the face of the choice that appears before us, for either we can let things work, or we can reject the proposed action. So what does the authenticity of this choice depend upon? A feeling?

Faced with a "feeling" that we are observing traces of our future, doesn't the authenticity of our choice then depend on our ability to identify such traces as having definitely been produced by us?

We have already pointed out the danger of trusting to traces of our own "folly".

For a long time, disturbed by these conclusions, I myself found it all rather hard to believe, the way we do when faced with an optical illusion, before someone explains to us how it really works. In reality, I took a long time to understand that it is time as we are usually made to conceive of it that produces this sort of optical illusion and stops us from feeling things correctly. The idea that time flows towards the future and carries us along, whereas in fact the opposite is true, is probably the main reason we are so handicapped when it comes to using time.

So how do we make our choice? How do we face up to the danger, in principle, of a choice not dictated by reason? Is it simply a question of staying detached after letting go, so that we can "let things work"?

In order to answer these questions, we will take a detour via physics and examine a fundamental function of time, one that allows the future to enter the present: the function of observation.

Chapter 8—Summary

To favour the realization of our intentions, we have to allow the current of information they are inducing in our future to "work", and this requires a certain amount of time: time for the water to trickle down.

Letting go induces a healthy attitude of mental detachment which prepares the moment when, as if by chance, the act takes place. The act happens automatically, as if realization of the intention had occurred before the act took place.

In reality, the Law of Converging Parts starts from the act which has already been realized and comes back down in time to find the path that returns the arrow to the archer's bowstring.

IX.

TO BE OR NOT TO BE

*In which we discover two meanings of life: to be, to act and receive
in the present, but also not to be, so as to allow the future
to look after our intentions.*

"Véronique, do you want to know the secret of the magic of time?"

"Are you joking? I don't even know what it's all about!"

"It's not my secret, even if I give that impression in my book. It is divulged in Gitta Mallasz's book *Talking With Angels* (18). She didn't invent the content, either, nor did Hanna either, a woman whose words she transcribed in her book."

"Oh, yes, I know that book, but the secret completely escapes me. So where do you think it comes from? Do you really think it was an angel?"

"I don't know, nobody does. Hanna simply says, "Be careful, these aren't my own words," but the person or people who were really talking never introduce themselves. Gitta calls them angels, or internal masters, or even beings of light, but even she isn't really sure, and anyway, it's not important."

"Why not?"

"Because the content of the words is enough for me. The rest is beyond me. I can give you an extract if you like."

"I'm hanging on your every word!"

"Listen, this is the Angel talking:

"I tell you a great secret: do not make plans with your head! Carry them out with your head! The plan is with the Father, all plans are. If you plan what you are going to do with your head, you are giving time free rein to realize it, for head and time are one!"

Then Gitta comments:

"Impossible to understand the meaning of these words. I clasp my head in my hands," and the Angel answers her:

"Do not fret: the Plan hovers over and above time. If you become one with the Plan, then you are never early and never late!" "

"I don't understand a word!"

"Nor did I at first, but much later, after doing all my experiments and even after starting to write my book, it suddenly became clear as daylight!"

"I'm listening!"

"Making plans inside our head risks preventing our intentions from hovering above time, that is in the future. For our head is already linked to the future, for head and time are one! So the head must busy itself with carrying out its work in the present, and not concern itself with the future, you see?"

"Hang on, that's it, we're back to letting go! So the head has to let things work, is that it?"

"Exactly, you see how clear it is? And when she says that the project is with the Father, that means that it is in his house, inscribed upon the Universe, so it will happen by itself if God so wishes! Do you know who God is?"

"Don't tell me you know that too!"

"No, no, don't worry. God can be whoever we want, but here he incarnates every possible thing, the land of the future, all of space and time. When you follow a path, it depends on the lie of the land, doesn't it?"

"I suppose so!"

"So there we are then. When our head concerns itself with the future, it's as if it were trying to dig the earth and not allowing the existing terrain to work. Moreover, concern about the plan will constantly modify your intentions, so how do you expect God not to give up on a plan which is always changing?"

"Well, yes, if you put it like that. So why does the Angel say: "never too early, never too late"?"

"Because the opportunities created by the Plan always come at just the right time when the ready-traced path towards your future reaches your observation!"

So the teachings from *Talking with Angels* confirm our conclusions from the previous chapters:

– our intentions might automatically "inform" our future (making our intentions rain down on it), possibly leading to a "signal" directed towards our present,

– if our head concerns itself with the present, our intentions will be able to perceive such a signal, more or less consciously but always at the right time.

I specify "more or less consciously", for the perception of the signal, as we have seen in "letting things work", is not always conscious and may show itself as a spontaneous impulse. We should not be surprised by the possibly unconscious nature of the signal we receive, for it often happens that we receive signals according to a process that is unconscious and apparently non-causal in origin: this is called intuition.

Could our intuition be the result of an unconscious decoding of information that has also come from the future?

Whatever the case, we will now see that there is no reason to be surprised, because the creation of reality itself, if we rely on physics, might be no more than information from the future entering into the present, or to put it simply: the future entering the present.

In this chapter, then, we are going to take a detour via physics in order to answer the following question:

How can our present really be formed from our potential futures?

This is the most fundamental question to be addressed in this work. It alone brings into play five premises which are very strange, yet all acceptable in modern physics:

- **the future is already realized**,
- **the future influences the present**,
- **the future is multiple**,
- **we possess authentic free will**,
- **our observation influences reality**.

Adopting all these premises boils down, no more, no less, to adopting the Theory of Double Causality, but to my knowledge, no physicist has adopted them all simultaneously. However, we will see that each of them is supported by great physicists.

The first one to suggest the first premise, "the future is already realized", was Albert Einstein, for this is a consequence of his Theory of Relativity. With it, he questioned our concept of time by assimilating time itself to a dimension, indissociable from the three spatial dimensions. In a letter addressed to the family of his great friend Michèle Besso, when his death was announced in March 1955, he wrote:

"The distinction between past, present and future is only a stubbornly persistent illusion. Time is not what it appears to be. It doesn't flow in a single direction, past and future are simultaneous."

Today, Einstein's concept is making a healthy comeback, with the so-called Grand Unification Theories, like String Theory or Loop Quantum Gravity Theory, which treat the universe like a block of space-time already deployed in the future as well as in the past, and where the present has no place. The second of these theories even claims that time does not exist.

We may well wonder: where do we get this impression of present time? Now, let us not forget that physics cares nothing for the phenomenon of consciousness, about which scientists in general know nothing. This being so, it is permissible to come out in favour of something obvious: the impression of present time is purely a property of consciousness and thus of observation. We will develop this point when studying the final premise, about observation.

The second premise, "the future influences the present", is not unthinkable within the framework of an already realized future, but it does, as we have already seen, come up against dogmas of causality and the arrow of time. In spite of this, numerous physicists, and by no means the most obscure among them, have taken the risk, as I do myself, of emitting theories implying a retrocausality, notably the French physicist Olivier Costa de Beauregard, who produced a hypothesis I will be returning to later. Here I will give a very recent example that made the headlines in the *New York Times*.

While I was writing this book, whose ideas must certainly have been in the air, two renowned physicists, Holger Nielsen and Masao Ninomiya, estimated in a series of publications that the failures in starting up the LHC, the world's largest proton collider, built in Geneva, might be a sort of "future warning": the future might be influencing the present to prevent us discovering Higg's boson, a particle actively sought by this accelerator. There could be a sort of unattainable future for some unknown reason. What we need to understand is not that these two physicists really believed in their own theory, but that they took advantage of the LHC's mishaps to present serious hypotheses in a light-hearted way, theories which would certainly have been rejected had they been put forward in any other context.

They had prepared the ground for this several years previously, when they published a first paper attempting to demonstrate that the irreversibility of time is impossible, a condition essential to their thesis. This publication went on to receive worldwide public acclaim in 2009, thanks to an astonishing coincidence: a bird carrying a piece of bread caused a short-circuit inside the electrical workings of the LHC, interrupting its cryogenic systems. The bird got away, but it lost its piece of bread!

For the first time, in the wake of this incident that corroborated the theses of Nielsen and Ninomiya, physicists began thinking about the existence of traces of the future. It has to be said that the event, rather incredible given the security measures in force at the LHC, came in the wake of other previous, very bizarre malfunctions, ones that had already led our two physicists to develop their bold theory.

The third premise: "our futures are multiple", dates from about fifty years ago, when the physicist Hugh Everett, backed up by Einstein's collaborator John Wheeler, introduced the concept of parallel universes in order to provide an extremely coherent interpretation of the problem of observation in quantum mechanics, as we will see later. Now this premise, too, is making a strong comeback with the success of the Maldacena's conjecture, which supports a ten dimensional theory of gravity in a holographic universe. The existence of parallel universes, whatever real or potential ones, is actually possible only through the addition of extra dimensions. Note that several theories agree on the need to add six dimensions, among which the String Theory and the more metaphysical one of this book, inspired by our own scientific works[4].

Everett explained that all the possible outcomes of an observation divided the universe into so many parallel branches. This has the advantage of solving the famous paradox of Schrödinger's cat which, locked inside a box containing a poison activated by a radioactive substance, has a one-in-two chance of being dead when we open the box. Now, according to quantum mechanics, as long as we do not open the box, the cat is at once dead and alive, and today this affirmation has become indisputable. Everett explains this by distinguishing one universe containing a dead cat from another containing a live cat. With extra dimensions, this poses no mathematical problem.

Today, multiple futures are opened by multiple doors, and the search for extra dimensions has even become official among the community of physicists using the LHC.

The fourth premise: "we possess authentic free will", still seems impossible to prove, for it is a metaphysical and philosophical question, but one shared by a large number of physicists for reasons of personal conviction. This is why, only recently, two renowned researchers, John Conway and Simon Kochen, had no qualms about shaking up the particle physics community

---

4. P. Guillemant et al, *A Discrete Classical Space Time Could Require 6 Extra Dimensions, Annals of Physics 388* (2018), 428-442.

by drawing the strangest conclusions from the overall results of quantum mechanics, in the form of the "Free Will Theorem".

First they defined the free will of an entity (a particle, a human being...) as a property according to which the state of the entity at any given moment cannot be described as the result of a mathematical function with a bearing on the state of the universe **before** this moment. Thus the entity is not entirely limited by its past and is partially free in its own evolution.

Conway and Kochen then went on to demonstrate that if we accept three axioms, one being causality, plus two other, highly realist axioms, in that they derive from the best-proven results in modern physics, axioms all quantum physicists consider to be true, then certain quantities measured by physicists, that depend on their choice of measurements, cannot preexist before these measurements occur, that is, they cannot derive from the information available in the universe and accessible before the experiment. From this they deduce their theorem: "if someone doing an experiment possesses free will, then the particles whose state the person is also measuring possess free will!"

What is surprising about this is that the theorem does indeed show a transfer of free will: if the particles' state is not determined by their previous history, then it is because of the free will of those observing.

We could deduce from this that free will purely and simply does not exist, considering that the idea of particles having free will is crazy, but this would not resolve the problem it poses for causality, and would, above all, be forgetting that the Free Will Theorem is in fact founded upon this very causality. Questioning it is exactly what allows us to get a better understanding of the theorem: when the observer's free will acts upon the future, the behaviour of the particles starts depending on this future through simple retrocausality, and in this case the "transfer of free will" is quite explainable and so appears only as an apparent transfer, an illusion: the particles are not free because they become determined by an imposed future.

So instead we should consider that if we take as axioms free will and the presence of the future, then Free Will Theorem constitutes a very nice demonstration of the Theory of Double Causality. Now all we need to find out is "how" our free will could influence our future!

We must not forget, however, that if, like Einstein, we admit the first premise, then the barrier of time no longer exists, rendering the future potentially accessible. But in what way? We do not know. Some think that because the brain emits electromagnetic waves—detectable outside the brain using electroencephalography—then the invisible space-time surrounding us might be able to record thought shapes. Without actually rejecting this possibility, I prefer to envisage an influence not confined to our four-dimensional space-time, in which extra "atemporal" dimensions could play a crucial role, but we will look at this later on. In any case, the possible influence of consciousness, or the power of intention, remains an open question.

We can, however, conceive the possibility of such influence in a humble way, if we consider nature's ability to cause all the complexity and organization of the living world to emerge, compared to the paucity of our own technology in constructing artificial humanoid-type systems, for example. So it is legitimate to wonder whether Mother Nature might be much better than us at mastering electromagnetic communication and everything it entails. If this is in fact the case, then nature's ability to synthesize organized electromagnetic information from living systems could be far superior to our technological potential in the field of mobile phones.

Whatever the means by which consciousness could influence the future, we will now see that this hypothesis is a long way from being far-fetched, for it has the advantage of solving the notorious problem of the observer in quantum mechanics.

The final premise: "our observation influences reality", has by now been adopted by a large majority of physicists in the world, but the problem of the "observer's status" in quantum mechanics is still a long way from being settled, for many physicists want to involve consciousness, and this provokes vigorous debate that creates a large divide within the scientific community, a debate that is still raging even now.

We all sense that the future enters our present by way of our consciousness and observations: these two things must therefore be closely linked. This appears to be all the more true today, in that physics, by banishing present time, has come to share the view that this present is closely linked to our consciousness, thus making it inseparable from our status of observer down here on earth. The problem here is that physics attributes an entirely fundamental role to the observer, that of imposing a unique, observed (measured) reality from among all the "observable" potentials, termed as entangled because they exist simultaneously in a multiple fashion before being measured. The way the observer might play this role, with or without his consciousness eventually intervening in the "choice of nature" contained in the observation's result, something that, for want of anything better, we attribute to pure chance, is something still hotly debated within the world of physics, for there exist no fewer than seven different interpretations.

The debate causing all the discord comes from the fact that nobody knows who chooses the result of the observation. Could it be chance? Or God? Could it be the observer himself? Or is everything determined in advance?

All we know is that at the moment of observation, one single reality is selected among the whole collection of all potential realities, because that is what we see at the microscopic level of a particle or molecule, where these potentials are described by a "wave function". Physicists have established that the observation of certain microscopic objects using sophisticated instruments could thus transform their wave function, which in essence contains multiple possibilities, into a single observed state. A unique version of our shared reality seems therefore to be imposed on our universe by each observer, who in some way becomes the "guarantee" for this uniqueness.

The crucial point in this debate, which allows double causality to confirm itself as a credible solution, is the fact that there are two distinct operations involved in the collapse of the wave function:

– the first is the collapse of state, which concerns the present,

– the second is the choice of the final state, which mainly concerns ... the future!

In order to avoid bringing consciousness into the first operation, decoherence theory, which enjoys a wide consensus among physicists, has been put forward to explain how collapse of state could happen naturally in the absence of an observer, by simple interaction with the environment. The problem with this theory, however, is that it does not explain the second operation, that of choice.

As far as the result of this choice is concerned, it is reputedly indeterminist —that is, produced by complete chance. Since the physicist Alain Aspect carried out his experiment in 1982, an experiment since confirmed many times, it has in fact been shown that there are no hidden variables which would allow us to claim determinist choice. Nevertheless, nothing forbids this choice from being determined before collapse, providing that it causes so-called "non local" hidden variables to intervene, that is, variables which operate "remotely" but in a non-causal way. There are only two alternatives for such remote, non-causal action: the first is to admit the principle of non-localness, which allows two particles to remain connected in spite of the distance between them, and the second is to admit the principle of retrocausality, which is based on the idea that the acting variables are hidden but are part of the future. In this second hypothesis of double causality, the choice would thus have already been made, but so long as the collapse of the wave function has not occurred either through observation and/or decoherence, it can continue to be refined in the future. This was precisely the hypothesis of the great physicist Oliver Costa de Beauregard, now deceased, who introduced it to the world several decades ago, although at the time it was considered esoteric. Today, that is no longer the case, for in the space of about thirty years, physicists have carried out and checked a number of experiments that defy causality, not the least of which is the so-called "delayed choice" experiment, in which it was noticed that the past could itself depend on our observation in the present.

Although the idea of an interaction by the consciousness upon the result of the choice was shared by such illustrious physicists as Eugène Wigner or Roger Penrose, it is still a hypothesis widely considered to be esoteric. I am in partial disagreement with this idea, because to my thinking, it stems from a confusion between consciousness and observation. That being said, in some special cases where present and future are included within a very short lapse of time, I do not exclude consciousness and observation having a joint, simultaneous role to play.

My objections to an interaction of consciousness in the present are the following: what happens in the case of multiple observation? Is the first observer in time privileged in determining the precise moment of collapse into a single state? If so, how many nanoseconds early? And what would happen if observation was not instantaneous but needed to be progressively revealed by a lengthy process? How does the degree of consciousness exert influence? At what moment will the choice of the particle's state occur? Is the wave function watching us observing it? Is there an "observation threshold" beyond which collapse into a single state is provoked?

So just what would a wild-eyed gaze produce?

Enough! All these questions, born of the conjecture that a single observer could simultaneously have the power to determine the state of objects—or the direction of events—and be able to select the reality that enters the present through observation, seem to me to make the idea that our consciousness may act upon our present scarcely credible. In this I agree with the majority of physicists, who nevertheless have to struggle against a widely held belief among various media and alternative movements, the

"New Age" movement in particular, according to which the observer creates reality. In my view, this is wrong because it is too reductive, the product of a naïve interpretation of quantum mechanics. Moreover, I am uncomfortable with the idea that the observer can choose in the present, when he has neither the means of his choice nor the consciousness of having a choice to make, nor even truly consciousness of anything at all, purely and simply.

The way that consciousness acts upon reality, which is something I do not deny, is actually more complex and more subtle. To understand it, first we have to dissociate all observation and physical action in the present from the direct influence of our free will in the future. By necessity, this influence requires consciousness in a way as yet unknown to us, but which the "presence of the future" allows us not to reject. This would give consciousness plenty of time to "engrave" its intentions.

When summed up in this way, the Theory of Double Causality solves the problems of decoherence theory, which supposes that a certain number of superposed states are capable, even before being observed, of getting disentangled all by themselves with a view to forming one reality, but without telling us which one will be chosen. Whereas as soon as the future is already prepared for the approach of the present, there is no longer a problem, for in this case we can conceive of all the alternative states which self-destruct during disentanglement as being other possibilities from the future, whose probability for observation is so reduced as to cancel them out. The entanglement phenomenon would therefore be a simple transition between a single reality and multiple potential realities inhabiting our parallel universes. All would be potentially "observable", but some would disappear in the future due to lack of observation.

Better still, however, double causality can even make everyone agree on the fundamental question: which one, out of chance, God, the observer and determinism, is responsible for the fateful choice?

According to the Theory of Double Causality, the answer is as follows:

– sometimes it's God, for if He exists we cannot exclude His taking direct action upon the future;

– sometimes it's chance, if for example the conflict between multiple realities is too fierce;

– sometimes it's the observer, when his intentions have correctly shaped his future;

– sometimes it's determinism, when the future depends on the sum of human intentions.

The morality is: everyone is right, and here we have something that can make the seven different interpretations of quantum mechanics compatible with each other, interpretations which, more or less markedly, have omitted to involve the future, something explained by the lack of models allowing us to formalize such notions as "letting it work" or "traces of the future".

To conclude, we can complete our diagram of the creation of traces of the future by distinguishing the following two phases :

– one phase for "letting the intention work", in which consciousness has to "let go" and stop thinking about the future, in order to come back into the present and participate in what's happening;

– one phase of "updating" which, through the act of observation, updates the arrival of a new reality or "branch" of life susceptible of making us turn off somewhere.

The act of "**letting it work**", which we can translate by letting go when it starts and by **detachment** when it lasts, would seem to be a very important process, because it is common to all situations depending on non-causal logic in which the Law of Converging Parts intervenes:

– letting determinism work, so that it will find a way to make me cross paths successively with different animals at the right moment, such that I won't get lost and will get safely back home;

– letting the archer's arrow work, so that it leaves the centre of the target and finds the coherent path back, taking into account every little gust of wind, allowing it to return to the bowstring and supply the fateful impulse;

– letting the rain work, as it falls in an area of multidimensional space-time and swells the streams, so that one of them springs forth at the right moment and delivers its message to the person who thought up the rain;

– letting the disentanglement phase of the process of decoherence work, which, starting from a point in a multiverse where all the possibilities of our Tree of Life are still in a state of probability, has to find a way to increase the probability of the path which concerns us and is likely to place itself in our field of observation.

Each time, then, it is an observation that is "passive" (in the Zen sense of the word), yet conscious of the present time, which validates the entry into reality of a non-causal potential, whether it be a coincidence, a piece of intuition, an unleashed arrow, etc.

Thus we see that consciousness does not have to intervene in the observation in a reductive way. It does not determine the observation, but merely broadens the field of possibilities prior to the observation which, alone, will then fix reality. I insist on the fact that consciousness does not intervene in the choice of the potential which becomes reality, and that only observation counts: there is no act of creation on the part of consciousness at the moment of observation.

Here, then, is the miracle of wave function collapse into a single state: consciousness contents itself with allowing the observation or not allowing it, it allows us to be live in the moment, or not to be and let the future act in favour of the realization of our intentions.

How, then, make a decision, faced with an observation that offers us several alternatives? Might not the real choice lie in choosing or not choosing to trust our intuition? Yes, but, then, what to do with an observation made by chance? For chance happens at the right moment when it is "caused" by second causality, and intuition could supposedly guide us. But is that always the case? Aren't we in danger of confusing the chance of second causality with a "false" chance?

How can we observe the chance which is bringing us back traces of our future or which makes us pull off master strokes, without running the risk of being wrong?

Let us look at how chance shows itself in the mechanism of second causality. Let us admit that a collective or individual potential of a human intention to "inform" the future has played a fundamental role in the fabrication of an event we observe by chance. It causes an opportunity for choice to occur. This is all very well, but is this future our own?

We may be uneasy when faced with an opportunity that seems to suggest a choice to us. Should we let our intuition speak? It would still have to provide us with all the elements for making a decision. When these elements are unconscious, everything is fine, we are prompted by an impulse which speaks for itself. But when it is a matter of conscious observation, of coincidence, what must we do? How can we interpret it?

If chance is the ally of the observation process that updates our life potentials, fine, but perhaps we are lacking some information to make our decision.

It is all very well saying that the universe has calculated everything to "pull it off" for us, but we may be missing an element that sways our decision!

This is what we will now examine, by studying the properties of chance.

Chapter 9—Summary

The Theory of Double Causality brings into play five strange premises of modern physics, all admissible, however, because they have already been put forward by famous physicists:

- the future already exists;
- the future influences the present;
- the future is multiple;
- we possess free will;
- our observation influences reality.

The mechanism of second causality is organized into two very distinct phases:

- a phase for "programming intention", during which the potentials favourable to the realization of our intentions see their future sources being watered;

- a phase for "letting intention work", which, having allowed the probabilities of an observable potential to increase, then brings this potential into the present by means of our observation.

This phase of observation may take place without any particular interference from consciousness. Consciousness acts directly on the future when we forge our intentions. But in order for this future to take root, we must let go, then stay detached. Our path towards this future is then prepared in coexistence with other potentials and the sole function of consciousness, along with observation, is merely "to be" in order to act and receive within the present.

This interpretation is one compatible with the principle of decoherence in modern physics: the result of observation is automatically prepared by the universe.

The question of the choice we may have to make after an observation is one that involves intuition or, failing that, information we still have to learn to detect.

X.

THE MAGIC OF CHANCE

*Where we realize that the Lord's ways may not be
as mysterious as people say.*

Irène does not often venture from home, but she is very happy in her remote house in the middle of the countryside, surrounded by animals wild and domestic with whom she likes to communicate. She even feeds a fox who keeps her hands full, quite literally, repairing the chicken-run he is forever trying to break into. Considering the special relationship she has with animals, I asked her whether she had any anecdotes for my book that would reveal their instincts and, in particular, their ability to detect our intentions.

"He's a wily one, my fox, you know, he can sense when I'm busy and sometimes he comes and fiddles with the chicken wire in broad daylight. In fact, it's often at midday, when I'm having lunch! The other day, though, I managed to get a photo of him on top of the wall."

"Good, but don't you have something obvious, something astonishing to tell me, something to make us wonder how he guesses? I don't know, maybe the fact you weren't keeping an eye out, for example?"

"No... But he does sense whenever I look out of my window. That's why I was so pleased to get a picture of him the other day! Every time I see him, he runs away!"

"That's not bad, but... Are there any times when it doesn't happen immediately...? When there's a short lapse of time between your absence... Ah, no, you couldn't tell anyway, you live by yourself."

"Oh, I see what you mean. You know, it's just the same with my cats when my friend Pauline comes to see me, she loves cats. They always know she's coming just before she arrives. Even when they can't hear her car, you know? Because of the bend in the road..."

"At Sorine? Right, so you mean they give you signs while she's still far away and her car can't be heard?"

"Oh yes, absolutely, when she must be over a kilometre away! Especially Lili, she comes and presses up against this window, here, and starts mewing and purring, and she only does that when Pauline comes. Why do you want to talk about that? It's just animal instinct, you know, nothing more than that!"

"All right, but instinct doesn't explain anything, it doesn't mean anything, it's just something we say when we don't really understand! But that's not at all bad, and I've read about that sort of thing before, in that book I lent you, you know?"

"You mean your theory offers another explanation than instinct?"

"Well, yes, it's because of observation. Instinct is explained by observation. The ability to observe is extremely important. Animals don't think or reason, and that means their ability to observe is much better than ours. Not just by using their eyes—they move their ears and detect scents. in fact, they detect lots of things and know how to recognize signs much more easily than we do."

"Signs? How can there be signs that Pauline's coming in what Lili observes?"

"Well, I'm not inside your cat's head. She makes her own codes. I'm talking about Lili's observation of unusual things that aren't all produced by your friend's arrival, but which for Lili signify that she's about to arrive. Because it comes from the future, do you see?"

"Umm, I'm not sure I'm with you there."

"I don't know, if Lili hears a sound far away, for example, maybe a bird or something, at the same time as she sees you getting up from your chair, or that a leaf falls, perhaps to her it means that Pauline is coming."

"But it's got nothing to do with it, I don't understand what you're saying, it's weird."

"No, it's not weird, quite the opposite in fact—it's very simple. Your Lili has independently made unconscious codes for herself that begin to operate when Pauline is in her car and getting near your house and thinking how enchanting the countryside is, and imagining the pleasure of telling you how beautiful it is, a few minutes later!"

"Goodness me, your story's verging on the magical now! Are you sure your readers won't think you've been smoking something funny?"

"Don't worry—actually on the contrary, in my book I spend the whole time boring them with rational explanations that are far from being so lightweight."

Among these not-so lightweight explanations are the ones taken from modern physics, that have helped us understand the two fundamental aspects of the mechanism of second causality:

– our intentions act in the future to prepare our potential futures;

– our observations act in the present to receive one of these futures.

When our observations speak directly to our subconscious, bringing forth some piece of intuition, we perceive information that orientates us towards the proper choice, without even knowing where it comes from. However, when these observations reach our consciousness directly, as with coincidences, then we need time to reflect, because we do not know how to interpret them. We do not know whether they carry meaning that actually concerns us, and even if we thought that were indeed the case, we still do not know how to decode the meaning and deduce some piece of information from it, an eventual choice or action. This, at least, is the case for the vast majority of us.

Here lies the nub of the problem about improbable events that happen by chance. Their very improbability makes us stop short, but without any instructions, we just do not know what to do with them. Some people see them as signs, but signs of what? We have already explored the rather dangerous consequences of such thinking.

Let us make a foray onto the Road of Time to try and improve our understanding of the tenuous relationship between chance and free will.

First of all, we must remember that it is impossible to realize an objective that benefits from chance if we follow our journey along the road without ever turning off in a non-conventional way, because then we would be behaving in the same conditioned way as the ordinary masses. We will also note that if we want it to be productive, then two conditions must be respected when coming upon a fork in the road that might suggest the right path to take:

– there has been a mental plan for a project (intention);

– discovery of the path has to be made by chance (observation).

Moreover, at that precise moment the intention has to have been forgotten, or at least not be interfering with our consciousness.

On the Road of Time, we generally have no other reasonable choice than to follow the road. And generally, even though we may from time to time come across an alternative route, a path or a stream that crosses our road, we do not feel concerned by any idea of turning off.

For the main road feels infinitely safer, and we need a very good reason for leaving it for another path which may lead who knows where.

On the other hand, if we have prepared ourselves mentally to change direction in order to find something, if we are ready to abandon the safety offered by the main road, if we are open to the idea of travelling alone, where no one, or hardly anyone, goes, taking a path that might have some surprises in store for us, then the situation is quite different. Everything becomes possible and we become receptive to clues.

But we must wait. For the moment we commit to this daring project is not the right moment for actually realizing our objective. And even when we allow ourselves to be distracted from our intention, the mere fact of happening on a path that suggests a possible turning is not sufficient reason for actually taking it. That way, every chance event along the way would interfere with our free will.

Contrary to what second causality would tend to make us think, the correct way to exercise our free will, therefore, is instead to make it as independent of such chance events as possible, the way we do in the causal framework when we give ourselves the means to realize our goals! No change here! For it is important that we do not spend our time watching out for opportunities. Not only would that be exhausting, but it would be akin to falling into the trap of the archer who uses reason when taking aim.

Let us now suppose that as we progress along the Road of Time, an animal crosses the road in front of us and disappears along a previously unsuspected path. We could find plenty of other examples, like hearing an animal emit an unfamiliar cry. What is important here is the simultaneity of some event already rare in itself, and its occurrence at the precise location of a potential turn-off.

You will say that this has nothing to do with our goal and that consequently, there would be no sense in our going on to take this path. It would be better to wait for another clue, connected to what we are looking for.

This is true, if we are indeed waiting for such a clue. However, if the fact of seeing the animal crossing at that spot awakens our dormant intention by making us smile inwardly at the idea of adopting this path, then there is scarcely any need to hesitate.

Yet we still have to be sensitive to this sort of "sign". Second causality cannot actually create a "temporal bridge" unless we are prepared to make use of such clues, otherwise no such bridge can be formed.

For it is our receptivity to these sorts of signs that in fact validates our using them, by virtue of a kind of retroaction, a temporal loop placed between our present and our already-present future. We could say that because the Law of Converging Parts knows we are receptive, this places more means at its disposal for sending us temporal bridges, for our openness to such clues is actually specified the moment we "programme" our intention. Thus we could "codify" the interpretation to be given to our observations, just like animal instinct! Except that here, it would be done consciously.

Now, isn't this already starting to sound a little bit magic?

We are beginning to catch a glimpse of how chance and free can work together, in this case when chance actually supplies us with a key.

This type of key, commonly called a "sign", is undecipherable only because we have never learned how to use it, still less programme the codes ourselves. This is proven by the fact that we cannot even make a distinction between the "sign" and its "code".

The code is the whole collection of clues containing the information we are waiting for, in this case an indication of the path.

The sign is the "signature", the "stamp" of the Law of Converging Parts that validates the authenticity of the temporal bridge: its improbability.

The key is the two as a whole—code plus sign.

So above all else, what characterizes a key is its signature: the very small probability of chance associated with it. If this probability is very small because there is a coincidence between elements that are in themselves improbable, then the code will come from the combination of these elements. Here, seeing an animal cross the road is improbable, but that is still not enough. It is the fact that this chance reveals the existence of a turn-off that makes it highly improbable, because animals can cross anywhere. When, in addition, it awakens an intention, then we have all the characteristics of a signature.

In real life, the Road of Time can be replaced by a train, an airport, a beach, the metro, a circus, etc. The path can be replaced by any sort of incident, however tiny, that we detect on one or other of these stages; it suggests an opportunity that might make us turn off, change something in our lives—if, of course, we feel it concerns us.

Now that we understand the possible alliance between chance and free will, we still have to define exactly what constitutes a chance event, and to know what chance we are talking about. For we are obviously not the source of all chance, and equally, some chances may only appear to be chances. Moreover, why do we so often invoke chance when confronted with situations where, quite on the contrary, our free will does not come into play?

In fact, we generally invoke chance in reference to an event that no one is responsible for. This can mean an event that might not have happened at all, or that might have happened differently. When we invoke chance, there is always an indeterminist element implicit.

For example, a tornado that destroys a house could well have missed it. A losing lottery ticket might have been a winning one. A wounded man might not have found himself in the path of a stray bullet. Someone out walking might not have found a banknote on the ground. Even though all this seems to occur independently of the will of the people involved, chance apparently has the power to complicate these people's Tree of Life.

In each case, every event was unpredictable and indeterminist: so at the same time there is an absence of meaning, there is also a very clear presence of turning points.

Could this mean that the richness of our Tree of Life is not due merely to our free will but also to chance? Could chance be competing once more with our free will in deploying our Tree of Life?

To answer this with a "yes" seems obvious at first glance, but it would, however, presuppose two things:

– the event that made us take a turning (i.e. bear or overcome something) is not dependent on our intentions;

– nor is this event the result of indeterminist chance.

In fact, we need to take into account the existence of determinist chance, which, as we have already seen, has no incidence on our Tree of Life because it intervenes only in circumstances where we are unaware of the true causes of events, events that are in some way inevitable.

Determinist chance is not real chance but only apparent chance: it is the result of our ignorance of the conditions presiding over the law of cause to effect, or as physicists say, our ignorance of the "hidden variables" that preside over the course of events. It corresponds to an event that is predictable, but incalculable, because we lack the elements to do so.

We will put to one side this sort of determinist chance inasmuch as it does not modify our Tree of Life; but it is important to know that these two sorts of chance are indistinguishable, and that this can be a source of error and confusion when we are looking out for signs or clues. Later, we will address the "joker" strategy that can be adopted when faced with this risk.

For now, we will isolate a subtle aspect of chance: how, at the moment of observation, it can suddenly change from a state of true, indeterminist chance to a state of perfectly determinist, false chance.

Just suppose that you have a lottery ticket in your pocket and that in theory, as long as you do not know the results of the lottery, you are potentially a winner. Is it a true chance when you discover, long after the draw, that you have either won, or lost again?

The answer is yes, but only on condition that you do not know what numbers you played, and neither did anyone else. For in this case it is clearly impossible to change your numbers. Meaning that if the result and your numbers have been observed, then what you discover is no longer the result of chance, or at least not true chance.

In the meantime, if someone else discovers your numbers then you necessarily become a loser or a winner without even being aware of it, while believing you may still be able to win; your future has been updated by an observation other than your own!

Because of this, some people who consider themselves unlucky often ask others to look at their ticket before telling them, as if trying to delegate their luck to them. So they have a sort of intuition about there being a link between the notion of chance and the notion of observation: it is the first observer of an event who "fixes" its content and prevents it from being modifiable. Beware, however: I did not say that an observer actually decided the content of an event. For we have no reason to think that there can be any correlation between what is seen and what is expected by the observer, and we even have good reasons for believing the opposite.

In physics, at the particle level, it is the "wave function collapse" that fixes the state of a system when it is observed. Could the same phenomenon happen at a macroscopic level? This is still a mystery, for at a macroscopic level we do not know how to identify the equivalent of *entanglement*, and it is highly probable that potentials are not entangled, but separated into parallel universes, which would explain why macroscopic indeterminism is so difficult to prove.

Let us admit that premise, however, and then come back to that quirk of indeterminist chance that suddenly disappears once an observation by someone turns it into a determinist chance.

Does this mean that, whatever happens, when we do not influence the event, then someone else has to have done so, and that in some way, chance does not exist, because it does not intervene in any choices?

Let us examine what physics teaches us about chance: based on our current scientific knowledge, we can deduce that there are two types of chance: on the one hand, macroscopic chance engendered by the "butterfly effect" of Chaos Theory, and more generally by the dispersive broken-glass effect, as well as many other phenomena (see Annexe), the sort of chance that presides over unpredictable phenomena on a macroscopic level; and on the other hand, a microscopic chance that operates at quantum level, labelled "wave function collapse", because it presides over the indeterminist behaviour of particles.

It is with the latter type of chance that we associate the notion of "updating" when the event it refers to is observed, "wave function collapse" or "updating" being equivalent notions. But there is no reason to think that macroscopic chance, whether chaotic or dispersive, should also not be the object of an update, since it too is just as much of an indeterminist chance, and this is precisely our basic argument.

The latter sort of chance, however, seems to be at work in various situations where our intentions sometimes appear passive, and sometimes active.

For example, whether a passer-by gets hit on the head by a roof tile or finds a bank-note on the ground is more the sort of chance where he is passive, whereas a lottery draw is more the sort of chance where we are active, a chance our "intuition" leads us to believe we could activate by luck; otherwise far fewer people would play the lottery.

From a physics point of view, however, there is no reason to believe we could trigger drawing the winning number any more easily than we could the falling of a tile, given equal probabilities. Which means that the famous chance that leads us to think we are passive before it, is no more or less updatable than the other, the one that is the object of games of chance. They are governed by probability. The two types are just as indeterminist as each other, there is no other difference between them.

In any case, according to chaos theory and quantum mechanics, both of which exclude any determinist interpretation where a single outcome is possible, indeterminist chance is extraordinary, because contrary to our usual way of thinking, it engenders the coexistence of an infinite wealth of potential futures, from which only one will be updated by becoming the one we will actually live. Would this not be to ignore the presence of the future, though, and forget the inexorable influence of our free will over the future?

So we could answer our question about chance by concluding that it does not exist after all, because everything that is not determined by the past is determined by the future, often long before it enters the present; but this would be playing down the possibility of the coexistence of multiple potential futures until the very last moment—the present.

Before we study this possibility, let us note that the coexistence of multiple potential futures due to macroscopic indeterminism obliges us to distinguish two forms of existence within one system, whether living or artificial, both of which are compatible with the laws of physics:

– multiple forms of possible but unobserved existences;

– a single form of lived existence, observed only within the scope of our past.

These two forms of existence are nonetheless real, insofar as space is big enough to hold them both.

And to be clear, from a scientific point of view nothing allows us to declare that any one of these forms is any more real than the other, outside our simple experience. I insist upon this difference between experience and reality. For if we were to claim that the reverse were true, it would be exactly like claiming that there were no other inhabitable planets apart from our own, on the pretext that we have never actually observed any. Or like claiming that Theopolis did not exist, on the pretext that we had not found it.

In this way, inappropriately introducing chance to elucidate the paradox of the observer in quantum physics is clarified by the spatial nature of time: all our possible existences really do already exist, but only one of them will be experienced. All we needed to do was make that distinction between "experience" and "existence". Nevertheless, the "reality" of both forms of existence is not to be doubted. The inappropriate part comes from attributing the function that ensures the passage of existence into experience, to chance. For realizing that transition is surely rather a function of consciousness, or at the very least, of observation?

The distinction between existence and experience is also the reason why it is not compulsory for us, with the model of our multiple-existence Tree of Life, to call on the notion of reincarnation, even though I have to be honest and make it clear that we cannot actually exclude it, either.

What I am about to propose may lead us to think that incarnation, or even reincarnation, may be of some use. What I mean by this is that, although the notion is still useless for our theory, it might not be useless as far as our universe is concerned.

Indeed, the universe does not wait for our future to become our present in order to structure it. It is already quite remarkable that the universe should accept the possibility that we might modify its structure through changes in our intentions. However, it is indispensable that these modifications be prepared without waiting for them to be updated by the present, so their stability can be assured. For it would be nonsensical for the universe to have to wait for my life to end, for example, in order to modify its structure all at once, by updating all the incidences engendered by my life. All the more nonsensical because these incidences are entangled with those of all the other people who do not die at the same time as me, and upon whom my existence has also had some influence, and so on.

It is more logical, then, for the universe instantly to update the whole of its future structure (and its past!) at each moment in the present time, this present merely signifying a simple process of updating changes already prepared, that is, mapped out in terms of their existence, but not yet in terms of their experience.

Within causal logic, that is, within second causality, we must reason therefore with the idea that changes in our real intentions instantly create a modification of our future existences—replacing an old one with a new one whose scenario has become more probable. In fact, we are obliged to work with probabilities, insofar as this new scenario is still susceptible to change as long as it is not updated.

We could, then, hypothesize that the change in the scenario, from old to new, happens when the probability of the new scenario comes to outweigh that of the old, at the very last minute!

Now, what happens at the last moment, just before a potential enters reality?

This is the moment observation comes into play! Now, we can easily imagine that, suddenly "whetted" by a warning or by heightened awareness, a capacity for observation could, at the last minute, succeed in favouring a new scenario's entry into reality, a new branch of our life whose probability was much lower prior to the warning.

This makes us see how complex the factors are that influence the entry of a new branch of life into reality, because they depend at the last moment on our capacity for observation. These probabilities oscillate!

The probability of a new life scenario being updated constantly varies, then, according to our actions and our mental state. It is obvious that if, for one reason or another, while we are already "engaged" in seizing some non-causal opportunity, we allow ourselves to be distracted at the last moment, then that scenario's probability will rapidly start to melt away like snow in the sun, or like stock-market values during a crash.

But if the old scenario, and all its other competitors, have at the same time also seen their probability plummet because our new behaviour has become incompatible with their realization, then the new scenario could see its chances of realization maintained in spite of everything, that is in spite of a probability that has fallen ridiculously low! Even if we are distracted!

This is when a phenomenon with very low probability will appear in our environment as if by magic—if, of course, we are prepared for it—and somehow wake us out of our distracted state!

Its purpose would be to maintain what is already programmed for us and becomes once again, at the last moment, the most probable version, if only in order to preserve the stability of the universe.

This is how, through a mechanism as rational as causal logic, the non-causal miracle of extraordinary chance comes about!

Let us now go even further and consider what can be the significance of such a mechanism, whereby an improbable event emerges because the causal route, the "reasonable" route, has become even more improbable. In which circumstances of life do we see just this sort of event?

Quite simply, when we take risks, or again, when we get lost!

Or, if one prefers, when we put ourselves into complex, often unforeseen and sometimes critical situations that force us to "get our feet wet" and put ourselves at risk, to become involved in a radical way that precludes us making any return to a life scenario made redundant by the very fact of our taking this risk. Examples abound: the risk of finding oneself in a very bad situation to help someone in trouble, of changing one's brand image by making a daring speech to one's co-workers, of facing up to a crisis situation, etc.

This is why it is a little hasty to conclude that chance does not exist, for in crisis situations, when anything can happen, no one knows which element will eventually determine the outcome, or rather, let us just say that it is too early to tell: we have not yet collected all the keys, the most important one is still missing. Before I start to reveal it, however, I would like to insist on the main, absolutely essential function of chance: that of **being good for our health**.

The process of non-causality must in fact be used in great moderation, for what we have here is a potential drug.

In order to guarantee our well-being, we must choose a different way of using chance than the one consisting of always waiting (without waiting) to see signs of it. For we can well imagine that in order to develop such receptivity, we would have to "vibrate" with a level of energy that would be quite incompatible with the tranquillity we all need in our everyday life.

However it shows itself, chance is still compatible with our Tree of Life, because it leaves us the possibility of letting it play instead of us, as if we were playing a joker, a sort of "I don't know" card that lets us withhold our interpretation or even not make one at all. When we do not want to "let it work", observe passively, feel, decode, then understand, then we have the perfectly healthy option of using our "joker".

This card should be played for as long as no new path imposes itself on us unequivocally, whether we update it or whether we are swept away on its current of conditioning.

For life is not a long and peaceful river, and our trees of life are all entangled together: even though we might update nothing at all, we can still suffer changes due to other people's updates.

Chance's "joker" also allows us to keep our free will and carry it over in time. If we never needed to put off decisions until we had more information, or a better game, then chance would have no need to exist and express itself using such a card. If we do not play our cards fully, then chance is held responsible for our helplessness, our irresponsibility, our bad luck or even for our destiny: in the latter case, the joker is played badly, or used too often. When played properly, it allows us on the contrary to put off a more effective deployment of our free will until the right moment. The everything happens as if, in exchange for our joker, we are called on to get ready, to "wake up": this is the true benefit of improbable chance.

Nothing prevents us from accumulating intuition and coincidence until such time as we decide we have enough to take appropriate action.

But take care: there is a saying that aptly sums up the need to be ready to assume the consequences of changes we risk triggering:

"Be careful what you wish for, you may receive it!"

How can we protect ourselves from such chance events, that are produced by the fact that second causality moves in such mysterious ways?

We have not actually addressed the qualitative question yet—that of good or bad. We have always implicitly supposed that second causality could only bring us good, inasmuch as it is meant to realize our intentions! But can we be sure of that?

It is time to look at the final key, the most fundamental property of the very essence of chance, and consider our Tree of Life once more.

How does a new branch become obvious each time a new bifurcation is available? Why, if we were heading along one branch, do we confidently change direction? How does change take place with serenity and in the absence of all uncertainty? How do we know whether, by turning off from a branch of our old life, we really will go towards the life we have chosen? How can we not only exercise our free will, but also find happiness in doing so?

For although we now know that chance may offer us an opportunity to go where our choices lead us, we still have no idea what factor intervenes in the quality of chance, in the interest it really holds for us, in the pertinency of its codes, in what gives it that tiny probability, in the ease with which it may be interpreted, and finally, in everything which goes to make it magic!

Maybe we simply need to water our Tree of Life?

So the factor would be liquid: to make new branches grow, we must water our Tree, otherwise the branches cannot grow and the forks cannot be born. Without water, no branch can create or encounter another, and so we risk taking paths along fragile branches. No flower can appear. Without water, not only can the Tree not grow, but when we follow its branches we cannot find any traces, for they are carried by water.

So what can be the real significance of all this water?

What is the vital liquid capable of watering the Road of Time and our Trees of Life?

What is the innermost essence of all these liquids, whose abundance increases the more able we are to create magic in our lives?

We will discover the answers to these questions in Part IV.

First, however, we must submit this theory to proper experimentation and test it against the facts, so that the immensely important answers we will subsequently be addressing can be authenticated!

Chapter 10 — Summary

Some coincidences become significant when they carry double information; a code and a signature. The code (or meaning) shows us the path to take, while the signature (or sign) shows us that the information is reliable, but only when what we observe is highly improbable.

The Theory of Double Causality is based on the fundamental distinction between existence and experience: things do not have to be observable in order to exist, but as long as things remain unobserved, they exist in every observable way possible.

The result of double causality is that nothing happens by chance, insofar as everything not caused by the past is caused by an unknown future. Chance without cause cannot exist, for the mere fact of observing it endows it with a cause that is hidden, because it is in the future. However, invoking chance is still excellent for our health.

This is because hidden causes may remain hazy, complex and mysterious, and it is important to remain sceptical and make liberal use of our joker and tell ourselves it's chance, I won't count that, I'll wait for more information.

In order for the meaning of chance to "derive naturally" and become obvious, in order for second causality to become pervious to water and produce "magic" effects, we must water our Tree of Life. But what with? That is the question!

PART III

COINCIDENCES

XI.

LAYING DOWN INTENTION

In which we discover an experimental approach to synchronicity that consists of talking to oneself while pretending to be someone else.

> "You only see the miracle coming
> If you forget yourself.
> That is the secret of all secrets."
> *Talking with Angels*,
> dialogue with Lili n° 19.

Sometimes strange coincidences, like so many tiny miracles, worm their way into our lives like an invitation to read a message meant for us, but most of the time they provoke little more than a smile or mild amusement, because we do not know where they come from or how to interpret them. Some people refer to them as "signs". But signs of what? Did the universe forget to give us the instructions?

So generally, we put them down to chance.

Many of us, convinced of our own common sense, or our own critical or scientific acumen, call "irrational" any attempt to interpret such "messages" that have no causal reality.

And yet, the phenomenon undeniably exists and has caused much ink to be spilled in the past (15) (21) (26), even though it remains an enigma to science to this very day.

Now, however, our Theory of Time gives us the keys to understanding the origins of many coincidences, especially the ones we call synchronicities, by explaining their mechanism and showing them to be quite natural phenomena, probably provoked by our intentions.

We have shown how our free will can make an ally of chance, bringing to our attention the keys to a virtuous circle of non-causal support for projects born of our intentions, as if they were "flowing from a source somewhere in our future". But we still need to know how to observe and decode the keys chance brings us, these traces of our future that have been generated by our intentions.

Without such keys, we are merely passive onlookers at phenomena of synchronicity that sometimes make our lives easier without our actually understanding by what "miracle" we could possibly have caused them. Unaware of their originating source, prisoners of determinist or rationalist dogma that attempts to have us believe that logic and "common sense" should be based on causality alone, we scarcely dare grasp their meaning or take advantage of them.

Knowing as we do, according to our Theory of Time, that causes can come from effects just as much as effects from causes, we are now led to consider chance as something other than a source of disorder. Since it is granted that chance can transport meaning in relation to our intentions, a new concept of the human race and our physical universe can begin to emerge, freeing us from the pseudoscientific, determinist paradigm in which we have been mired for several centuries.

It is difficult to take the whole measure of this new concept of time, for it is so very far beyond us. Nevertheless, we must be very careful when we confront the breach it opens, for it is quite capable of making us talk and act utter nonsense. All things considered, I have an inkling that it could lead us to some sort of spiritual elevation, which makes me think that it is about time that scientists in general—and not just the physicists who opened up the breach long ago—began showing an interest in questions of "faith" by reconciling science and religion, metaphysics* and spirituality, conscience and soul.

If this book were to serve just one useful purpose, then I would want it to contribute to understanding the fundamental principle at work in all creation and all projects whatever they may be, including science and technology, for this is the foundation stone of double causality, of the reunification of science and religion and the union of science and spirituality.

Before I began to write this book, I had gathered enough personal experience, thoughts and reading to be convinced of the reality of a phenomenon unexplained by science. I wanted to rise to the challenge, never realizing quite how far it would lead me! But I had no proof, and felt a pressing need to find an experimental approach to synchronicity.

It was while reading *Talking with Angels* by Gitta Mallasz (18) one day on holiday in Haute-Provence that it all clicked and I had that "Eureka!" moment that led me to a satisfactory experimental protocol.

I had already been mulling over the idea that synchronicities could be provoked voluntarily. Reading that book allowed me to understand the rest of the "rules" or attitudes that had to be respected for there to be a chance of triggering the desired effect. It just so happened that I was travelling along the Road of Time with a companion who tended to encourage me to make these sorts of experiments. I learned rather late in the day that it was certainly thanks to her support and especially her love that I recorded some very impressive results. The protocol I followed would certainly be deemed inadmissible for any experimental approach that wished to be taken seriously, and so it was only valid for me, personally, but I can't resist the pleasure of telling you my story of *"Talking with Angels"*, however intellectually private it might be.

In order to understand this experimental protocol, first let us see what we know about the phenomenon of synchronicity, independently of any theories. It is a well-known fact that most of the time, we produce these manifestations quite unconsciously, without any idea what conditions we would need to fulfil in order to produce them consciously.

In a generally passive way, some of us are even quite used to regularly experiencing coincidences. They seem to occur naturally and be more or less related to moments of soul-searching, while at the same time being completely involuntary. The most inconvenient of them can be covered by the expression "Murphy's Law*", but I will not venture any further into this pseudo-law that merely bears witness to our ignorance of the true mechanism at work.

We have all experienced synchronicities in rather funny ways: you are thinking of a friend and just then she phones you, or the astonishing variation on the same theme: you pick up the phone and find yourself talking to the person you were just about to call. On another level, you meet someone on a train and you discover you have a mutual friend! There is a common saying that aptly sums up the effect such odd happenings have on us: *"It's a small world!"*

The first principle of the protocol I conceived for "provoking" such coincidences is based on respecting the apparently indispensable delay we have remarked on between the instant we intentionally "programme" our future and the moment chance brings us a signal in reply, symbolized by a bifurcation we could take on our Tree of Life.

Its second principle depends on the question: could this bifurcation be trying to teach us something?

Let me remind you what this bifurcation means: this is the moment a trace of the future worms its way into our present like a signal, or an invitation to let ourselves be swept along by coincidence, as it brings us information that could help us make a choice. So it can indeed be trying to teach us something, just so long as whatever it is remains quite distinct from the choice itself.

What is difficult about trying voluntarily to provoke synchronicity is all contained within the first principle, and mostly concerns the waiting: when does it end? And what should we do during this indispensable period of letting go? How can we recognize, without risking being mistaken, any sort of sign that indicates the end of the waiting—in other words, a synchronicity?

We can see that within this process, everything happens as if we were proceeding by somehow depositing a piece of information and then forgetting about it, but only temporarily. We are waiting for something, without dwelling on it too much. But how do we make this deposit? Should we talk to ourselves? Should we say a phrase out loud, to imprint it firmly in the air? How can we make this impression, whose aim is to inform our future?

The protocol I adopted was quite simply the following: I had to invoke an intermediary entity who played the role of a receptacle for my intention. Such an invocation had the advantage of making it infinitely simpler to let go and wait, because the entity acted as guardian of my intention.

Any clear definition of an acceptable way of representing such an entity has yet to be made, and this is something we will examine later on, but in the end, it can very well remain abstract or even mathematical, supposing we have a proper model of our conscience: what we are doing, in fact, is activating properties produced by a sort of timelessness of our conscience, or at least from one part of it, if we can put it like that.

Now, to understand exactly what answer we expect from this sort of invocation, we need to take a close look at the difference between coincidence and synchronicity. The experimental protocol we have just suggested is only suitable for synchronicities, not mere coincidences.

Synchronicity is coincidence loaded with meaning. However, sometimes a coincidence can be thought of as a synchronicity we do not understand. It may not actually be meant for us personally. So we will only classify as synchronicities coincidences with obvious meaning. Numerous examples of this are given in *Coincidences: Chance or Destiny* by Jean Moisset and Michel Granger (6), a book that teems with experiences of both types and draws a clear distinction between them.

However, once this type of synchronism starts playing in your favour, it becomes impossible to continue talking about coincidence. For example, you need to find somewhere to park in the centre of town where everything's full, and you find a space right next to your destination just as you get there. Some people, myself included, are "luckier" than others in this respect, and I admit that it is a very practical tendency to have, although it is not systematic.

Life coincidences may take a large variety of forms. For example, some numbers may repeatedly cross your path as if you attracted them. Personally I have seen a great deal of this, and this bizarre aspect of my own experience that I will be describing later on (The double 22) is very instructive for understanding the power of second causality, a power that can even go as far as to "recreate the past" whenever traces are absent or actually lost.

But whereas a coincidence is simply a strange curiosity we often wonder what to do with, a synchronicity is particularly surprising, for it "immediately" appears loaded with meaning for the person experiencing it, as if it contained a message meant for them.

The most classic example of true synchronicity is the one given by the father of the concept himself, the famous psychiatrist Carl Gustav Jung. This is the Golden Scarab story: while one of his patients was undergoing analysis and describing to him a dream in which she had met a Golden Scarab, at that very moment the self-same insect flew straight against the window of the consulting room, leaving both of them deeply disturbed. The meaning of this fortuitous coincidence, according to Jung, was that the disturbance it caused served to boost his patient's stagnating therapy. And indeed, it turned out to be a major turning point in his patient's life, putting her on the road to recovery. However, we may well ask ourselves where the intention that triggered it actually came from—the patient, or her doctor?

As a general rule, the most commonly cited examples of synchronicities are associated with events likely to be life-changing for the person they happen to. Take the example of people who often travel by plane and are delayed by exceptional circumstances that cause them to miss their flight, on the very day the plane crashes! It would appear that either they did a very good job of informing their future of their intention to stay alive, or others had done so for them!

However, there is nothing quite like experiencing things for oneself: the most meaningful synchronicities I ever encountered were the ones I experienced voluntarily on the Road of Time, and their meaning was linked to my project for writing this book.

All the time I was working on my project for this book, I was accompanied by numerous synchronicities, all bringing grist to the mill for my project, harvested by "invocation" along the road, until it was almost as if the harvest gradually became something I systematically expected to occur, and that actually made the synchronicities more likely to happen. A whole series of coincidences experienced over a ten-year period had prepared me for this sort of incomprehensible phenomena. They were all related to the double 22 (chapter XV), but I had never really felt these numerical coincidences carry any real meaning. So for me, they had not been synchronicities. I could never quite understand what they might signify, apart from the fact they always happened on the most important days of my life. They imposed themselves on me without my asking them to.

Before I devote the greatest portion of Part III to them, let us take one example I experienced quite independently of all the others: the day before I went to spend the weekend in a small guesthouse, as always on this very same road, I met a geo-biophysicist, or water diviner if you prefer, someone who specialised in identifying geopathic interference zones in houses, frequently due to water seeping through underground.

Having myself studied geo-biology ten years previously, I talked to him about his subject with good-natured scepticism, bringing up the irrational nature of certain practices using a divining rod. I wasn't questioning the credibility of his methods for detecting subterranean water, but rather the so-called correcting of any pathogenic effects using devices supposedly emitting "frequencies"! This particular water diviner, a very pleasant man indeed, ended up offering to analyse my own house for me, but since I was intending to move I declined the offer and asked him if he would be prepared to analyse somewhere I envisaged buying instead. At the time, I was thinking of restoring a ruin on a piece of land situated somewhere along the Road of Time.

The next day, dazzled by the reflection of the sun on the road, I had a minor car accident, bumping into a rock and leaving my car stranded at the garage for several days, forcing me to go home in a taxi. It was during my conversation with the taxi driver, who was well acquainted with the area and its inhabitants—and an attractive lady, besides—that I learned that a number of people had thought about buying the ruin I was interested in, but nobody had ever actually gone through with it because apparently there was water infiltrating the ruin's foundations.

To cut a long story short, in the same weekend I learned quite by chance about the problem of the water beneath the house, just after meeting a specialist in detecting subterranean springs. What is so surprising is that the meeting occurred before anyone told me about the problem, which then came up completely independently!

It was not until later that I understood how odd this was, when, having constructed my entire theory, it seemed obvious to me that the information about my ruined property had reached me in an entirely natural and non-causal manner, after I formed the intention to recruit the water diviner to look for springs.

Let us return to our project of invoking an entity to serve as a depositary for our intention. This strategy came to me as I thought about the conditions under which synchronicities generally occur, and after having researched the question in several books, when I realized that the common denominator for these conditions was mental preoccupation. The more vital and therefore more loaded with emotional "energy" this preoccupation happened to be, the more likely it seemed to me to trigger the phenomenon. And yet it also seemed, according to other examples, that the fact of adopting an entirely relaxed and totally unpreoccupied attitude, also had the power to trigger the same effects.

What I gathered from this, drawing a comparison with the technique of letting go such as in the secret art of archery, was that an unconscious "deposit" or an emotional discharge, born of some preoccupation or other and subsequently released, must have taken place. The apparent contradiction could be explained thus: in the first case, we consider the subject prior to the "deposit", while in the second, we consider the subject after he or she has been liberated from it.

During the process of making a "preoccupation deposit", whatever it might be, confiding it to a friend, an animal, a plant or the universe, the ensuing state of mental release seems to improve its efficacy. And here, preoccupation clearly rhymes with intention. It would even seem that what is actually deposited is somehow the "preoccupying energy" of the intention, and that a feeling of confidence subsequently fills the mind of the subject.

In his book *Necessary Chances*, Jean-François Vézina (31) presents a thorough justification of the interpretation whereby synchronicities occur in situations where we are preoccupied by the need to find answers at certain moments in our lives, especially when we are open-minded enough to receive them. Without going so far as to describe the possibility that we might systematically provoke them, the author does, however, give a perfect explanation of their message function and the conditions necessary for obtaining it. I quote: "The psyche has to be disturbed... Moreover, the message has to be a very important one for our own development."

And this is how I was led to use a "Guardian Angel" to play the role of the depositary. In the absence of something that really exists, simple faith in this entity leads to a salutary feeling of confidence and an indispensable state of letting go. The Angel represents an abstract entity who plays the role of our own double, to whom we express our intentions, our problems and all sorts of questions. It is an internal dialogue that aims to find answers, but by really entering into the spirit of the Angel game, we allow ourselves to receive these answers in an a-causal* manner, by carrying out the function of deposit followed by that salutary letting go which brings confidence.

What this comes down to is a bit like organizing a meeting with ourselves in our own future, by exploiting the possibility of dialogue with this "double" who already lives there! When, on the contrary, we persist in our solitude, entertaining intentions we don't know how to express outwardly, whether it be with acts or words, all we do is close our minds to any dialogue that would help prepare our future, because all it needs is to be informed of our preoccupations!

Have you noticed, for example, that frequently, when we desire something, especially when it is a question of obsessional desire, the thing tends to happen once we have forgotten it, or even once we have stopped wanting it!

This shows that desire is not an effective way of creating the right opportunities, probably because it imprisons the object of desire by preventing it from being transformed into future "programming energy". It also shows that we intuitively know our desire has a role to play: could it be to contain energy and keep it in the present while we wait for the process that will memorize it in our potential futures? We can see that this role is initiatory and that it is important not to remain at this initial stage of desire, so as not to block the subsequent stages that allow it to be realized: these being the intention for a real encounter with the object of desire, confidence in the effective memorization of the intention, and then the letting go and mental void. Whence the expression: don't mix dreams and reality!

Of the subsequent stages, mental release or retreat is fundamental. The mind is a parasite that tends to calculate everything, something that merely serves to "project" according to the law of cause to effect, whereas it is precisely this law we must reject if we want to make more effective use of our free will using non-causal logic.

By judiciously building on double causality, we can affect the future while knowing that the present derives from the future and not just from the past. This amounts to **a retroactive loop mechanism**. The concept, however, is somewhat abstract, and the best thing to do is experience it as an internal dialogue with one's future.

Nevertheless, we are in fact talking about a real deployment of a retroactive mental loop destined to produce the answers we are waiting for in the shape of "signs" for us to decode, when we know that we ourselves set the rules of the code!

We could reduce the phenomenon to a simple "mind echo", where the echo is somehow reflected back to us by a zone in our future. But would it be a mere echo when that echo could actually bring us extra information?

Whatever the case may be, we now have to write the rules of the code.

Chapter 11 — Summary

Synchronicities or meaningful coincidences usually happen to us in a completely involuntary way.

Starting from the hypothesis that they may be a result of our intentions, we suggest that it should be possible to trigger synchronicities by depositing an intention in the future.

Often, such a deposit would be done unconsciously, and conversely, it could be entirely consciously blocked by desire. Conscience, then, does not appear to be a favourable factor, which confirms the necessity for letting go.

In order to make this deposit, we suggest liberating our conscience by entrusting our request to a depositary entity, whose main role would be to take up our intention and allow our conscience to let go.

XII.

THE MODEL OF THE SPIRIT

In which we discover that the evangelical words "Ask, and ye shall receive" and "God helps those who help themselves" are two complementary methods for increasing intention's effects on the future.

> "Do not believe that anything is impossible!
> The law of weight is: the possible.
> The law of the new is: the impossible."
> *Talking with Angels*
> Dialogue 31 with Hanna's Angel

The entity for dialogue that acts as a receptacle for our intentions can be interpreted in two ways: we can think of it as a "Spirit*", as in the expression, "Spirit, are you there?", or as an "Angel", referring to an even more abstract entity we still call our "Guardian Angel", one entirely likely to be confused with our own selves. However, in literature this guardian Angel is always assimilated to a spirit, which is why we will call our experimental model for internal dialogue the "model of the Spirit".

These two forms of spirit allow us to decline our theory of double causality according to two versions:

– a minimalist version that considers there to be no spirit incarnate in our life or even on our Tree of Life, and that what we commonly call "spirit" does not exist, except to represent the part of us which lies hidden inside the extra, invisible dimensions of space. I would let myself call this an "Angel";

– a conjectural version that pushes the potential consequences of our theory about our concept of the universe to their utmost limits, and which might, most notably, to explain this universe's evolution within eternity, call on notions of "incarnation" or "reincarnation" of spirits.

As I have already said, I am not going to adopt a position on this in this book, and will instead be content to support the minimalist theory, with a version of the Angel that can thus be practically reduced to an abstraction, or even a mathematical model.

However, I would not want to stop certain readers taking the liberty of going further, in the absence of any proof to the contrary. Meanwhile, since the Angel is itself often considered as a Spirit, it seems appropriate to me to explore this notion of Spirit, defined as an entity dissociated from the body and entitled to a concrete existence, even though it may be on the other side!

Whatever the case, in order to enable the power of realizing our intentions, it is always recommended, when using this internal dialogue, to take this entity (Angel or Spirit) seriously, by always considering it as distinct from oneself.

This is because, by really giving us the impression that we are making our request to someone else, it makes it easier for us to act as transmitters. So this receptive entity will somehow do for us what we cannot imagine doing ourselves in our future.

In any case, the Spirit model supposes that we admit that there is a part of our conscience distinct from our brain, at once complementary and immaterial, in other words, something that could not be the result of the billions of neuronal interactions that are the expression of our thoughts and acts.

In the thirty years I have been interested in synchronicity, I have always thought that the phenomenon was a product of the duality of our existence, making a very clear distinction between conscience and matter. It is only very recently that I have understood that there is no need to distinguish conscience from matter, but only to distinguish our present experience from our reality outside of time. But there was one nagging problem: is it necessary that this part of our conscience we call "spirit" or "soul" be distinct from our brain and from everything generally pertaining to matter?

Which adds up to the following question: do we need a concept of a disincarnated spirit?

It is the advances in modern physics that have led many physicists researching this apparent duality of the universe to look for what might be the scientific reasons for the different natures of what, on the one hand, we call the physical world and matter, animated by the laws of causality, and on the other we call "conscience", "spirit" or "soul", animated by other laws we have yet to discover.

Among the theories in physics that might support this duality, quantum mechanics is systematically invoked, even to the point of exaggeration, but so is what I would somehow term its "macroscopic counterpart" chaos theory, though much more seldom, because of the indeterminism it, too, generates. But there are other theories, such as the famous String Theory (7), that are extremely interesting on a mathematical level for grasping these questions, because they attribute many more dimensions to the universe than the four we are capable of seeing. String Theory has not yet been verified, and is even under dispute, but its addition of six extra dimensions is a recurrent result from several independent works, among which our own contribution. However, what is important is how coherent it is with the laws of physics, and the fact that its hyper-dimensional nature is not what is being called into question. In Part IV, I will be developing the reasons why, thanks to these theories, we have no need of a concept of dis-incarnated spirit for our spirit model.

I will now give the simplest possible definition for the entity I call Spirit or Angel: a spirit may be considered an extension of our own "individuality" in those dimensions of our universe imperceptible or invisible to us, beyond the four we know. Add to this that the extension is necessarily atemporal, or, in other words, exists outside of time, because it represents the part of our conscience that instantly modifies the fields of probability in the potential futures on our Tree of Life.

Knowing, as we do, that we are enclosed within a universe that has more than four dimensions, maybe ten or eleven, in fact, according to String Theory, then there necessarily has to exist some sort of prolongation of our being that characterizes us in the six complementary dimensions. It has to be the case, because it is a simple question of mathematics: we have to exist with an equal number of dimensions as those in the space in which we are immersed, even if we could possibly almost not exist in some of them, being no more than a single point, for example. Yet this is still a real point, one that actually exists, and what we have to note is that there is no reason to think that this compulsory extension of ourselves shouldn't have its own characteristics of conscience too, in which case we would be talking about an undetectable, imperceptible part of our conscience, or in a word: a "spirit", as far as we can imagine it to be.

To make it easier for us to imagine, I cannot insist too strongly on the necessity of understanding the pre-existence of our future, as well as that of our past. If only we can succeed in imagining our existence out of time by means of the fact that we could, for example, act upon our future right now, at this very moment, using the simple power of our thoughts, then we will be home and dry, our thought will have grasped the concept of spirit!

And maybe I am even more accurate than I think, because if, reading the last sentence, you have just understood this notion of spirit for the first time, then this will probably be the first of a set of modifications of your behaviour and your intentions that will trigger real movements of your spirit outside of present time, thereby helping you to grasp its true nature.

When we experience physical death, we think that we cease to exist, at least in the four known dimensions, because our body disintegrates and we can no longer feel anything. But we know nothing of what happens to the part of us that was living in our future in the other six dimensions[5] !

In the absence of knowledge, we can always suggest hypotheses, but let me make it clear that there is no need, for the purposes of our argument, to conceive that the spirit I am talking about might survive such disintegration. What may well be left behind is that part of us that is quite alive today but of which we have no awareness in the other dimensions.

It would be unfair to qualify our approach to the spirit as irrational, because it aims to be experimental and is based upon coherent hypotheses. However, it would be accurate to label it as "spiritualist" or "spiritual", the two concepts being related, but, it has to be admitted, each quite as intangible as the other. This being so, our approach implies a belief in certain things but in a way that is correctly founded on the fact that this faith plays a fundamental role in producing phenomena: it is a fairly well-known fact owing its origins to the interaction of our own conscience in the actual experience itself. Somehow it is at once an observer and observed. It is impossible to do anything but have faith, unless one displaces or delegates the problem to other observer-subjects, such as mediums, for instance.

However, wanting to experiment on an interaction with a spirit, implying a search for communication stamped with a certain faith, does not mean that we are not allowed to maintain a certain healthy level of doubt. To help me put together this experiment, I first set about reading the most fully documented works dealing with the question of spirits (11) and angels (9), but also religion, with the intention of finding a common thread that would allow me to approach the phenomenon of synchronicity from another angle: its ability to be reproduced.

From this point of view, it would certainly have been embarrassing if doing such experiments had meant my having to be initiated into the rites of table turning, trans-communication or any other means of paranormal

---

5. In more recent books and publications, we have proposed that three extra dimensions are associated with our self-consciousness and the three other ones with our "future being" or higher self.

communication, whether they implied the eventual intervention of a medium or not.

Talking with Angels, on the contrary, came across as a more subtle experimental method, by which I mean with a much weaker level of energetic stimulation, even to the point of being difficult to measure or evaluate, and with results much more liable to be reproduced.

Having researched Angels in several different works, some of which included very interesting personal accounts (9), but having failed to find anything about how to create a dialogue to try to produce immediate results, I put together the following synthesis: the means of communication was of little importance, the only thing that really counted was the state of mind—or possibly the modified state of conscience—that I needed for opening up an adequate channel of communication with my future. Once this was done, a terribly simple and particularly "gentle" means of stimulation naturally imposed itself: **Just put an inner question directly to the Angel and ask him to provoke a synchronicity to express his answer!**

Why not? This would seem to suggest that one is addressing a truly dis-incarnated being, when in fact that is not the case at all! Knowing that we are dealing with an intimate part of ourselves that inhabits our future, then aren't we best off putting it to use to make our lives simpler?

Let us note that this request for direct "manipulation" of the future from a starting point in the present is based on the decision to interpret improbable phenomena that might subsequently face us in the shape of answers. Meanwhile, in a most curious fashion, this is what contributes to making them more probable, thanks to retroaction. By requesting a display of synchronicity, I am acting in the present on a potential future that I am, quite rightly, going to consider being implicated in my request, because it will effectively influence my subsequent path. In this way, I am significantly increasing the probabilities of every potential scenario likely to bring me traces of the future I have programmed for myself.

In my request for a dialogue, I had no need to know at the moment I put my question whether I had been heard or not! I had a rendezvous with time. I simply gave the Angel a day to reply to me in the immediate future, and even an extra day if he wished to give me another answer.

Traditionally, communicating with an angel is meant to bring help to the person who asks for it. Many people commonly use this method in everyday life, even if only through the intermediary of prayer, in order to obtain what they are seeking, such as help for someone else, for example: people think of asking for support for a family member or a loved one, and some even light a candle to signify their request quite clearly.

Having finally merely rediscovered the virtues of prayer, I concluded after a little reading that the sincerity of the request, a truly open mind, genuine faith or sufficient spiritual preparation, full of confidence, love and compassion, constituted the necessary elements for a successful operation.

I added these elements to all the clues gathered from our Theory of Time:
– creation of non-causal order;
– intervention of indeterminist chance;
– intentions expressed without knowing how to realize them;
– new intentions;
– risk-taking;
– letting go.

To all these, we had to add the need to break with the habit, or leave the beaten track, during the phase for harvesting answers.

It was reading Gitta Mallasz's book, *Talking with Angels*, that provided me with the most important elements regarding the "spiritual preparation" and particularly the mental rules that had to be observed, among them *the authenticity of the request, respecting free will* and *developing the "inner smile"*. Gitta's book is unique in two ways: the dialogues have an "experimental" connotation derived from the Angel's availability and the reproducibility of the meetings. Having personally evaluated the responses of Gitta's Angel as being original and powerful (transcribed by a medium?), I was partly inspired by this book. The question as to whether the original author of the texts really was an angel, if it was all real or just imaginary, seemed to me, frankly, to be incidental. What was important to me was learning to catalyse a phenomenon liable to demand a modified, or at least unaccustomed, state of conscience.

The main directive I drew from this reading is the following: "Ask, and you shall receive an answer!"

What could be more obvious? For the needs of the experiment that concerns us, however, I extended this principle to the following phrase, our experimental hypothesis:

"Find an inner question for me and you shall receive an answer through the intermediary of signs you will have to interpret, signs that will be confirmed by synchronicities I will provoke."

In order to give myself the best possible chance of success by asking my questions the right way, I wrote down the eight following rules for mental preparation:

1. To have a genuine need for help.
2. To make a request related to something that truly preoccupies us when we make it.
3. To take the risk of "committing oneself" through dangerous, unreasonable and above all unreasoned behaviour.
4. To ask for something whose realization would have a real incidence on our path of life (new intentions).
5. To maintain our free will: and certainly not ask the Angel to choose in our stead.
6. To attain a sufficient level of detachment and letting go.
7. To see a genuine inner smile grow inside us.
8. To break with the habit and stray from the beaten track at the moment of making the request.

However, these attitudes should be thought of as "technical" elements that may well have no effect whatever unless a ninth and fundamental element is introduced, one we will see emerge in the final part of this book, so that we respect the thread of our reasoning.

All these "rules" concerning our mental—or spiritual—attitude can be considered as different phases of the "cycle of second causality" that we will be approaching together with this ninth element.

As to the advice to "stray from the beaten track", it is obvious that if we never "budge" from our chair or if we remain imprisoned inside our habits, there is no chance of our making any unusual, meaningful encounters; and so it is not worth making any request.

This is why the actual intention to provoke a situation that will allow an answer must be present at the moment of making the request. This corresponds to the famous words:

"God helps those who help themselves,"

a phrase to be employed just after this one:

"Ask and ye shall receive".

In Gitta's book, the simple word "Ask!" is pronounced a good hundred times by the Angel, as if this sort of invocation were a fundamental rule signifying that no answer and no lesson can be transmitted to us without our first asking for it.

And so I began this novel experience with inner dialogue as I spent a few days on holiday with my partner Danièle, while we explored the territory of the geological reserve, starting from the southernmost point of the Road of Time.

Chapter 12—Summary

The depositary entity of intention is described as an Angel or Spirit, the part of our conscience that remains timeless, that resides in our future and reflects our intentions.

The fact that it really exists stems from the omnipresence of our future and our capacity for modifying the probabilities of its potentials. In some way, the Angel incarnates this timeless upheaval.

In order to call on it, or, in other words, energize the potentials it encompasses, it is necessary to respect the right attitude, which can be summed up in eight points: a real need for help, genuine preoccupation, risk-taking, the seriousness of the request, free will, letting go, an inner smile and conditions that favour the unpredictable.

A ninth, essential element will be examined after we have gone into this in more depth.

XIII.

TALKING WITH ANGELS

*In which we will find it hard to believe that
the Angel is simply a part of us.*

"I will speak of the smile. The mouth on the human face corresponds to matter: it is below. The mouth is pulled down by descending forces and raised up by ascending ones. Every animal can moan and cry. Only the human can smile. That is the key. Smile not only when you are in good humour! May your smile be creative, not artificial: smile creatively! If gravity acts, it blocks here. Everything is dragged down, everything! The mouth belongs to the earth and gravity belongs to the earth.

The smile is an image of deliverance, a symbol. Creative force raises matter.

How could you find your path if not by smiling? I dwell in the smile and I am your measure. The smile symbolizes mastery over matter. If you read a book, you hold it close enough to read it easily. If you want to read me, come close to me: I dwell in the smile...

I, too, speak of the smile. You pass it by, so familiar is it. You are unaware of what it means. The smile is a bridge over the old abyss. Between the animal and what is beyond the animal, there is a deep abyss. The smile is the bridge. Not the laugh, not the grin: the smile. Laughing is the opposite of weeping. The smile has no opposite...

The key to all your acts, to all your teaching, is the smile. Try it! Smile! See whether your pupils can find the inner smile! Then they will move differently. The smile accomplishes more than any gymnastics...

A smile is the prayer of every tiny cell, of everyone... and it rises up to here... it rises above all else. To smile is so simple! And yet no one is aware of it."

(*Talking with Angels*, Dialogue 35)

Travelling along a deliberately improvised itinerary, Danièle and I passed through countryside of immense beauty. The amazement it inspired in us probably helped encourage the occurrence of surprising synchronicities I experienced during our few days' holiday. I can add another positive aspect, in the same vein: even though Danièle shared with me a level of confidence much less intimate than the one I shared with my Angel, her love and acceptance contributed even more assuredly to the experiment's success.

What started out as the "Angel game" soon became a fascinating relationship. Because let me say right away, the main reason that compelled me to keep making requests of him was this: it seemed to me that, unless I had very special powers, the thing that replied to me so clearly had to be distinct from myself. It was not until later, once I had understood by what mechanism the thing I had considered an Angel could derive from the properties of time itself, that I could admit that this Angel might be nothing more than an echo of my mind, a reflection of my intentions. However, I am still not sure of this, due to an annoying tendency always to doubt everything, including a theory desperately trying to pass itself off as a minimalist, if not to say simplistic, model of time.

My first request of the Angel was a complex one, for I was genuinely preoccupied (2nd rule) by exactly how I was going to set about a project of a book on time, a book I had not even started to write yet. I was on holiday, free from professional cares, and Gitta's book fascinated me. Reading it was like an external impulse pushing me to conduct my experiment, for at least two good reasons: on the one hand, I had just worked out the mental attitudes that had to be respected, and on the other, I wanted to make the most of my holidays, which would soon be over. But I wondered how I could reconcile a reasoned approach with such attitudes, especially the need to produce an inner smile. To make my task even more complicated, a phrase from Gitta's angel had me deep in thought, threatening to call into question the entire feasibility of my project:

"Instinct in humans has been contaminated by too much intellect."

This statement caused me to doubt, because this is how I interpreted it: to seek knowledge, in this case the truth about time, by carrying out experiments that called on an instinctive communication with spiritual entities—wasn't this stupid, senseless, or even simply forbidden? Surely knowing everything was by necessity incompatible with angelical communication, which called upon instinctive resources, entirely outside all field of knowledge? Should not instinct and reason remain distinct, separate, incompatible, precisely because instinct is needed when reason is lacking?

What must be understood is that at the time, I had no mind model as yet. I still believed in an obligatory duality separating two worlds, the world of consciousness and the world of matter, and I had only a vague sort of intuition regarding double causality.

As this doubt turned into a genuine preoccupation, I decided, perhaps out of derision, perhaps out of facility, to make the following request of the Angel:

"You know I have doubts about the feasibility of a rational approach to your world. I feel as if I am touching the unknowable, the impalpable, the forbidden! Shed light on this question for me today, using any sign you want to!"

I pronounced these words in thought only, to avoid worrying Danièle.

Because of the 5th rule, I could not ask a question needing a yes or no answer, because this would have led to a choice on my part. Whereas I had to bear in mind one of the Angel's comments that frequently recurred in my current reading, and which was, perhaps, one of the most important passages in the rules of the game:

"The human being keeps its free will and choice."

Having asked my question, saying it inside me clearly enough so that anyone reading my thoughts could understand, I tried gazing at the sunlight through my car windscreen (because angels are beings of light, aren't they?), and wondered what sort of signs I might receive in reply while merely contemplating superb scenery, but then I remembered what Gitta's Angel had said:

"If you truly listen, even the stones will speak!"

And in fact, the only things around me were stone and rocks and mountains! Thanks to this affirmation, I had faith in my request, and indeed, it was the simultaneous nature of this faith and the absurdly rational side of it that made me give an inward smile, while helping me take a healthy and detached step back: I was ready!

Relaxed and amused, I thus recalled the function of the inward smile I had discovered in Gitta's book, a prerequisite for any communication with the Angel to be effective, and I supposed that I must be on the right path. I do admit, though, that in time I inevitably lost a little faith and began to worry about my own mental state, too, something that did at least have the benefit of making me forget my question that much more quickly as we crossed the countryside ... until I made an encounter, less than an hour later.

As I walked around a labyrinthine village discovered en route, we were just drawing near a little church when I noticed a door standing wide open, leading to a strange sort of tribune, decorated with posters aimed at tourists. Straight away, just opposite the entrance, a photo of St Teresa of Lisieux set caught my eye. I was surprised by the piercing gaze of this woman who seemed to be talking to me, as if the photo were alive! At once, the question I had asked the Angel came back to me. In order better to understand my surprise, it helps to remember that St Teresa was not only a saint but also a writer, a theologist and author of the well-known poem called "To my guardian angel". This simple fact made my encounter with St Teresa begin to look like a "synchronicity", since I was just then trying to begin an experiment whose goal was to call up my own guardian angel! However, I was in a church, and there was nothing so very surprising about coming across angelical signs here. So this nod from St Teresa still seemed insufficient for me to draw any sort of conclusion from it.

However, as I got nearer to the photo to try and read the quotation written below, I was stupefied to discover a phrase from St Teresa that lifted all my remaining doubts on the subject. The quotation immediately made me realize that I was looking at an answer here, because of the strangest contrast between the religious character represented by St Teresa and this statement:

"All I have ever sought is Truth!"

The reason for my surprise was that this "search for Truth" was precisely the central theme of my recent question to the Angel. Curiously, here it was coming from a saint who might well have made us ask: why the word "Truth" instead of "Love"? When did saints start seeking the Truth? Why, on the very morning I doubted the Angel's ability to understand a quest for scientific truth, had I stumbled on this phrase from an angelical person recommending such a search for truth?

Was this synchronicity, though, or just a simple projection? If it was, then did its improbability lie in the link between St Teresa and angels, or in the contrast between the quotation and what we might have expected to read there, such as "All I ever worked for was Love" or anything else relating to Love?

Coming from a saint, I could not imagine a search for Truth, because to me, the paths to Truth and Love were quite distinct and possibly even incompatible. So I was fascinated by the quotation, and in the end I realized that the mere fact of my being so amazed was justification enough to tell me that I was in fact looking at an answer to my question.

Although I tried to force myself to sustain my doubt, I concluded that through the intervention of St Teresa, the Angel was telling me that there was no reason not to search for truth on matters of spiritual or religious faith. Which was exactly the answer to what had been preoccupying me when I asked my question.

This, therefore, was clear encouragement to pursue my quest for dialogue. The only thing that had been bothering me about the quest had been swept aside: the apparent incompatibility, at first glance, between my excessively rational experimental attitude and the receptivity apparently necessary for contacting such an angelic entity.

Scepticism returns with a vengeance, however, as soon as one becomes a little tired; and so it wasn't long until I started telling myself that the improbable nature of an encounter with someone like St Teresa, in a church, was actually fairly low, and that the quotation's originality, especially its affinity with the contents of my own question, only seemed improbable: they were simply the product of my imagination, meaning that because of this the encounter could not really qualify as a synchronicity, much less an answer from the Angel. As a rationalist, all I could do was look for reasons to doubt, by calculating the probabilities to see whether I was really looking at a synchronicity, but such probabilities were terribly hard to calculate.

Since I was on holiday, after all, and couldn't spend forever pondering St Teresa's quotation, I decided to take a photo of the quotation and put off my conclusions until later. Nonetheless, all these doubts had me wondering whether St Teresa's quest for the Truth had been scientific or spiritual, and what the word "Truth" actually meant for her. After all, maybe St Teresa could not have cared less for any sort of scientific truth.

Thus I asked myself about the notion of Truth in the spiritual domain, tormented by the question of compatibility between two types of "knowledge", one spiritual and one scientific. Perhaps I was misleading myself in my desire to find some sort of experimental or scientific truth about the Angel's world, and perhaps the only truth its answer could make me lean towards was a moral one, centred around the authenticity of being, and orientated towards love. So I didn't really form a question, and my thoughts remained centred around the initial question.

The next day, we came to Digne and, as is my wont every time I have the chance, I nipped into a bookstore to look for anything new in the "Science" and "Spirituality" sections. I immediately noticed a book called Sounds of the Universe. Intrigued by the title, I picked it up to see what it was about, and was surprised to read the subheading:

"The Link between Soul and Science."

Now, only the day before, doubting the interpretation I should give to St Teresa's quotation, I had wondered about the compatibility between spiritual truth and scientific knowledge. I immediately thought, well, well! Another coincidence! Does the link between my present preoccupation and this subheading show that the Angel is giving me another answer? I stood speechless, looking at the book, before asking myself another question: was the Angel suggesting I read the book, or was its entire answer contained in subheading, the answer being that there is indeed a link between soul and science?

Rapidly leafing through the book, I was disappointed by its contents, too abstract for my taste, endlessly beating about the bush without ever actually grasping the question in any original manner. It was more of a compilation of analysis about the relationship between Soul and Science than any synthetic body of thought. Disappointed, my scepticism came back to the fore.

I put the book back and started looking for others in the same section. Just at that moment, before I even had time to find another book, the bookseller placed a copy of *Talking with Angels* right there on the shelf, the book I myself was currently reading!

I was so astonished that I could not keep back a small exclamation, as all the tension left me like someone who has just been completely disarmed and can no longer fight or harbour any doubts. I thought:

"But what is he saying to me?"

And instantly the answers came pouring in: he was telling me in the subheading of the first book that I had grasped the fact that there was indeed a link between soul and science, but by showing me *Talking with Angels* he was making it clear that if I wanted to find out the truth about this relationship, then the only book I should be reading was the very same one already sitting on my bedside table.

I had just received a lesson from a "Spirit"! And it coincided perfectly with what St Teresa had already taught me: worthy representative of a spiritual world though she was, all she was doing was looking for Truth! And so the nail was driven home.

By putting *Talking with Angels* back on the shelf, the Angel was therefore telling me:

"Stop right there, you're about to go off course, don't forget my book!"

Why this "Stop right there"? To make me understand quite clearly, through intuitive thought, the answer to my quest for new reading matter: if I wanted to avoid straying off course, I should stop looking for any other books before I had finished reading *Talking with Angels*! The Angel was simply reminding me that I would find my answers, not in this bookstore, but in the book already in my possession.

Out of excessive doubt, however, I did not continue reading the book, and it was only two years later, after I had finished the first version of my book, that I discovered that the secret of double causality was actually to be found in Gitta's book. If I had only bothered to keep reading it back then, I would probably have saved myself a great deal of time.

However, this second synchronicity at last gave me faith in the reality of an existing dialogue, whatever the entity may be. At the time, it seemed to me that its answers could be summed up in the following message:

The Angel's answer:

"Forget your search for a new book and concentrate on the one you are already reading, *Talking with Angels*. Just remember that I am, a second time, encouraging you to seek a link between soul and science using your experiments with me in synchronicity. To prove it, I gave you the bookseller's synchronized action."

Back then, I was a long way from imagining that I would go so far as to propose an interpretation of the soul (see Part IV)!

In any case, it was clear that I had just experienced a perfect example of synchronicity, at once highly improbable and totally pertinent to my own questions.

Realizing this, however, was not designed to reassure me about how I should proceed if I wanted to pursue the experiment. It was too powerful. It would have been much simpler to sustain my doubts and come to the conclusion that the dialogue was not working. Everything would have gone back to normal and I would have carried on with my holiday, my mind at rest.

I mean, can you imagine facing the consequences of such a realization? Me looking out of a gaping threshold onto a path of deviancy, like a child to whom no one has given any limits about asocial behaviour! Surely psychological imbalance is just around the corner?

What, indeed, was to be done in the face of such a situation? Carry on with the experiment? Ask new questions? Wrack my brain trying to imagine how the next synchronicity might show itself? Orchestrate travels to new places or provoke encounters in order to produce a new answer? How can one not think that every time we cross the threshold into places favourable to new encounters, our Angel might use them to send us a message? And if he doesn't, how can we avoid being disappointed? How should we compose each new question? And how can we then forget it to make an answer more likely to come? How can we avoid wondering, each time we have a choice between two roads, which one the invisible entity is making us lean towards in order to reply to our request? Why, in fact, should there not be a better path? Does the Angel have the means to reply whichever path we take?

It's enough to drive you completely mad!

The last question was the only one I could give a clear answer to: no, the Angel cannot answer me if I take a known or familiar path, because there is a very strong chance that this will destroy any possibility for dialogue (rule n° 8). In order to provoke an answer from the Angel, one must create novelty (God helps those who help themselves)!

The problem is that systematically creating novelty while waiting for a message is very embarrassing. Like a good tourist driving his car along the path of new and unexpected experiences, one ends up unable to resist provoking novelty by systematically leaving the beaten track, and it is very easy to get lost this way.

It was at this point that, to get rid of my tendency to interpret every single sign as a message, I had the idea of entrusting the following instructions to the Angel:

"When the clock on my car reaches 191,191 km, a number corresponding to my children's dates of birth, send me some signs to let me know what I should do with them next week, and where you suggest I take them."

My children were in fact meant to be coming to stay when I got back from holiday, and I wanted to take the opportunity of organizing some activities we could do together, but I hadn't yet got anything sorted.

Aside from my writing project, this was my main preoccupation. By formulating it like a rendezvous centred on the mileage, it had the advantage of stopping me from constantly watching out for signs of an answer from the Angel to show themselves. All I had to do was wait until I reached the prearranged mileage. Nonetheless, I couldn't resist the temptation to calculate the approximate location at which it would take place. Having already asked the Angel my question, and not being taken in by this anticipation which ran the risk of making me try to influence the answer somewhat, I clarified my question:

"Please arrange it so that, even though I'll be inclined to anticipate where it's going to happen, I'll still be convinced by your answer. I do find it hard to stop myself intervening—by making my journey longer—to reach the prearranged mileage somewhere I find particularly pleasant, the place I would like to live with my children, for example. So you will have to use your imagination so that I can't calculate it like that and so I can still pick up your message and know it really does come from you. So you will have to prepare a lovely surprise for me when I get there."

I didn't say these words out loud, but I firmly expressed my request inside my head.

I did actually have a vague idea about approximately where I would prefer to be when I reached that mileage: on the mountain to the south-west of Digne. I wanted excellent "energy" to be present. I wasn't familiar with the mountain, but I wanted to avoid reaching 191,191 km on the motorway on the way home or on a shopping estate. I couldn't really see myself busy taking photographs of the surroundings on the motorway or at a supermarket, with the idea of looking at them later for signs of a positive response.

This way, I made the Angel's task more complicated to avoid his having to disappoint me, because I still had my doubts. At worst, I would be rid of him. Looking back, I could see that the synchronicity in the bookstore had not been enough for me. I wondered whether the Angel was really encouraging me to go on reading that famous book. If so, was it to encourage me to continue to pursue my experiment, was it really to encourage me to write my own book, or was it simply to broaden my personal knowledge of the Angel himself?

As tiredness came over me, I even began doubting whether or not I really should write a book about this method for generating synchronicities, seeing how awkward it would be for me to talk about the Angel. Quite evidently, he was not just an abstraction anymore; indeed on the contrary, he was looking more and more like a real entity. As a scientist, I didn't want to risk making a fool of myself. And at the same time, my project for a book was starting to take a nosedive.

I had over 50 km to idle away in order to reach the desired mileage on the mountain I had chosen, and it started to make me excessively restless. This is why I decided to take Danièle out to a restaurant and how what I have already described in the beginning of part two happened to us: two synchronicities, one in Digne, one in La Javie, coming less than 24 hours after each other, clearly bearing the same significance, so much so that my companion, experiencing them just like me, exclaimed each time:

"That means you simply have to write your book!"

So the order was repeated twice, as if to put an end to my doubts, while the project for a book remained a part of my plans. However, the doubts I harboured made me risk making too little of everything I had just experienced. For this to stop, I had to experience a whole avalanche of synchronicities bearing the same significance: I had to write on a precise subject, and I needed to be firmly determined to bring a term to a dubious attitude that at first glance was in danger of making me abandon my project.

There is an interesting comment to make about this sort of avalanche: scepticism doesn't necessarily hinder the production of synchronicities, so long as it doesn't detract from the open-mindedness needed for observing them. On the contrary, in fact, it would even seem to become an ally in perfecting one's determination.

So scepticism displays two apparently contradictory aspects: it can be considered as an absence of faith, but it can also be seen to appear as a doubt that helps reinforce that faith!

In effect, let us take a faith acquired too easily and confront it with an interpretation potentially subject to error. In this case, we can say we are no longer in the presence of faith, but rather of devotion, or even naïveté. What would such faith be worth? It is obvious that any ill-founded faith that loses credibility risks diminishing the probabilities of updating any future that might be in accordance with it.

We can conclude from this that a certain dose of scepticism can actually reinforce the power of intention, and that it may even contribute to favouring a healthy sort of detachment.

All in all, then, I had experienced four synchronicities that finally succeeded in convincing me of the necessity of setting to work on this book, especially since the last one was accompanied by the observation of a double 22, as if this were a signal that my hard-won determination would bring about a change in my life. I will be going further into this question of the double 22, which also deserves its own chapter.

Using four synchronicities, the Angel had thus twice confirmed the subject, and twice reminded me to get down to work. When I look back on this experience, I remember it as if I had received an order!

An order I apparently gave myself by passing along the corridors of times in my future! Strange echo, isn't it?

I still have to relate the final synchronicity we experienced just before the end of our holidays, the one concerning that famous mileage, 191,191.

Having left La Javie and driven some thirty kilometres, I stopped the car at the exact designated mileage (191,191.0) south-west of Digne, one or two kilometres from the village of Champtercier, on a narrow road in the foothills about 700 m up, still on the Road of Time. At first glance, there was nothing remarkable to be seen, apart from superb scenery, as ever. I wondered what I should read into the scenery and even took a few photos, when opposite us, about 100 m along the road, I noticed a turn-off with a sign saying "Fontliesse Campsite".

A campsite! I had actually been thinking of taking my children camping the following week, but an uncertain weather forecast and a cold month of August had led me to envisage a different programme that I had as yet to decide.

My first thoughts were, well, it's obviously going to be nice and so I can take them to camp, which will be perfect! And I started calculating the probability of coming across a campsite somewhere along this kilometre: if I counted a few dozen campsites along several thousand kilometres of roads, then that gave me a probability of about 1%: not bad, but did it deserve the genuine AOC of "true synchronicity"? Chance may have had a part to play, too.

Since it seemed obvious to me that I had to choose the turn-off to the campsite—without the signpost, I would have carried straight on and the car would have stayed in the middle of nowhere all along kilometre 191,191—we stopped three hundred metres along, opposite the proprietors' house, while below the house the campsite itself lay on the slopes, apparently quite empty of any occupying tents.

Curiously, all the windows of the house were flung wide open, inviting us to go in. Hardly were we inside than we realized that the house also served as a gîte, but that all the beds were completely empty and unmade: apparently no occupants, something confirmed by the proprietors we had woken up.

After noticing the empty beds, I suppressed my inner interpretation: would my children's beds remain empty, too?

All in all, the house was particularly welcoming, open, charming and situated in the sort of scenery one dreams about. I would have loved to live there. In the main hall downstairs, there was a poster on the wall with a quotation from Jean Giono, of which I will quote a few lines:

"The man who wishes to enter these happy lands will find its doors open to him. Haute-Provence is above all else the home of light and silence. It is possible to walk for days on end, quite alone, and experience incomparable joy, order, tranquillity and peace. This land teaches you about the noblest organization on earth: simplicity filled with wisdom makes its joys the most peaceable and enduring: it surrounds you with such dazzling logic that from then on, you are inhabited by a god of purity and light."

These words made me shiver with joy, and it seemed obvious that the Angel was answering my request in a magisterial way. The warm welcome we received from the house's inhabitants confirmed this, a couple with whom we swiftly made friends as they showed us around their gîte.

I had a feeling, then, that the Angel had answered me, and that he could hardly have done any better, given the exceptionally harmonious atmosphere of the place, but as yet I didn't quite understand his message, which only replied to part of my question:

"What should I do with my children?" Take them to a campsite, fine... but what sort of activities should I plan with them?

I only understood the true meaning of this synchronicity a few days later, when, having waited patiently for them at the airport, I realized that my children were not coming to spend their week's holiday with me: seeing the absence of tents at the gîte and all those empty beds, my companion and I had had the same thought, one I had immediately pushed aside. Despite the deep disappointment I felt, the Angel had replied to me in a pertinent manner and what was more, with a great deal of compassion on this occasion.

As soon as I got back from holiday, I sat down to the task of writing this book.

Chapter 13—Summary

An avalanche of synchronicities purposely generated by "asking the Angel questions" is at the origin of this book.

XIV.

LOGIC WARS

In which we will try everything to save causality, except at the cost of making it irrational by continuing to hide its twin sister.

In this chapter, we will look at the different kinds of possible explanations for these synchronicities, according to the logic of simple and double causality—the latter, after all, being no more than an extension of the former.

Our first attempt to explain these phenomena in a causal way can be justified by the fact that it is important to do everything possible to try to save that good old logic, "common sense".

To save this causal logic, we will "strike hard" by enriching it with concepts from quantum mechanics, such as the observer paradox. At first glance, there is no reason why this logic should be of any more advantage to non-causal logic than to its colleague, which is why we will use it to put our Theory of Time to a more serious sort of test. The supposed intervention of the observer's conscience in the result of the observation does in fact have powerful consequences, because it is liable to provide us with unexpected explanations for our experience.

For now, let us forget our Theory of Time, and ask ourselves the questions I asked myself after this little adventure, before I had a full grasp of the mechanisms that form the subject of this book.

How do all these coincidences happen to occur at the right time, the right place and be synchronized with what I am expecting from them? Why are they so instantly significant? How is it that they are systematically carried by some event that is improbable, in the statistical sense of the word?

We could see signs everywhere, which would be a pathological attitude to have. But we must not get confused: signs accompanied by improbable events are different from others. If, for example, I am watching clouds pass and I make out a shape that inspires me to make some interpretation, if I see a sign in it, then there is nothing synchronous about that, no element to validate that sign; it is just a mental projection.

Synchronicities, however, appear to accompany the interpretation lent to them, like a stamp and a signature being placed on a document to validate its authenticity.

The strength of causal logic is to make chance intervene while implying that we are not truly capable of evaluating how improbable events may be, because they depend on factors that are known and conscious, on the one hand, and unknown or unconscious on the other. Their improbability would thus be an illusion resulting from the fact that we are not conscious of all the factors that influence phenomena.

Just to recapitulate: during the four days I was on holiday, it is hard to say exactly how many coincidences I experienced, because some were grouped into a single event. However, I can break them down into nine phases, even if they have very unequal probabilities of occurring.

First of all, there was an initial series of six coincidences in reply to my questions about my project for a book:

1. The photograph of St Teresa, author of the poem "To my guardian angel", one hour after I had submitted my request to an "Angel".
On first analysis, I think we could invoke chance here, because when you enter a church, you can expect to encounter angelical signs, and when you visit a village, you often visit the church.

2. St Teresa's quotation: "All I have ever sought is Truth!", after a request about the search for truth.
My first analysis here too is that chance can be invoked, because of the way the quotation corresponds to my request of the Angel, because there cannot be too many well-known quotations from St Teresa.

3. The book on the link between soul and science, in reply to my question about the compatibility between the search for scientific and spiritual truth.
Here, too, we must invoke chance, because I was looking in the "spiritual" section, so there can be nothing so very astonishing about finding books there dealing with "soul".

4. The perfectly synchronized way *Talking with Angels* was replaced on the shelf, coinciding with the moment I was looking for another book to inspire my quest, when in fact I was already reading it.

Stop! There comes a point when you have to stop invoking chance because it ceases to be credible, the probabilities here are so very small: on the one hand, the synchronization of the gesture, and on the other, the synchronization of the identity of the book with the main theme of my chosen subject.

Clearly, if we invoke chance here, we are in danger of missing something. We will get locked into an "at first glance" sort of approach and single-mindedness.

5. The only free parking space right outside the bookstore interpreted as a sign to encourage me to write the book.

Upon first analysis, I have to invoke chance, because it might be dubious to lend any interpretation to it. I clearly "projected" a reply to a question I had been asking myself onto this coincidence, after the fact.

6. The restaurant, "The new novel", that finally banished my doubts on the subject, just when I was getting ready to return from my holiday.

This is quite funny, because it confirms the previous interpretation, but we can still invoke chance if we consider 6 as separate from 5. What is disturbing is the presence of both. When in doubt, however, I carry on invoking chance. By acting this way, I again risk being single-minded, but it is for a good "cause", if we can put it like that.

To sum up, the only event I retain as not being attributable to chance is 4. The problem, though, is that if I retain 4, then I have fresh grounds to call into question the way I have placed 1, 2, 3, 5 and 6 in the category of chance, because these coincidences are related to 4 on several levels: on the question of angels, and on the question of my project for a book.

So until I can find a better way of doubting them, I have no choice but to reconsider all my coincidences, because of this link between the most improbable of them and all the others.

7. A campsite and a gîte, at the exact spot where my mileage reached 191,191 km, when there is a very high possibility of ending up somewhere in the middle of nowhere or outside a closed-shuttered house.

Here, it is a delicate matter to invoke chance, because the probability of coming across such a fitting place with regards to my question was extremely low: maybe as low as one in a hundred, which gives us an idea, although chance is still a possibility. I will risk it.

8. The empty beds in the gîte and the absence of tents on the campsite, in reply to a request on my part assuming my children would be there the following week, when in fact they would turn out to be absent.
It is normal for beds to be empty in the middle of the day or at the end of the season. What was rather unusual was the doors all being open, so that we could see for ourselves. After all, everyone is free to do as they please. Let us put that down to chance.

9. And lastly, the exceptional nature of the gîte: its originality, harmony, charm, the magnificent quotation from Giono, all the doors of the house flung wide open, the superb view and the warm welcome we got from its occupants, woken from their afternoon snooze, all of this corresponded perfectly to my request to find positive vibrations in the sort of place I would like to live.

STOP! In every respect, finding somewhere that corresponded perfectly with my request was obvious proof. The probability of finding such a place was even smaller than simply chancing on a gîte or campsite. It is absolutely incontrovertible that the probability of this must be close to one in a thousand.

So I will also retain 9 as an event "probably" not due to chance. But for the same reason as 4, if I retain 9 then I have to reconsider my analysis of 7 and 8 in terms of chance, for both are related to 9 and somehow reinforce it, just as 1, 2, 3, 5 and 6 reinforce 4.

And so I have to rehabilitate all my coincidences.

In conclusion, I am eliminating chance as a "causal" factor and so I have to reconsider absolutely all the events from 1 to 9, and try to find another type of causal explanation.

If it is not due to chance, then we have to admit that these events are correlated in one way or another to my interpretations, even if those interpretations are partly mistaken. This is not the nub of the problem: with this hypothesis, I accept that there may exist a correlation between my thoughts and the course of events.

This is the moment we have to extract some irrefutable piece of logic and rescue causality! Let us begin by analysing 4 and 9, the strangest events of all: supposing that there is a causal explanation for 4: then perhaps we should deduce that the bookseller unconsciously waited until I happened to be standing in front of one of his displays to put *Talking with Angels* back on the shelf. That would mean information had come from me and "circulated" in the air, leading the bookseller to carry out this action, as if she were in some sort of semi-hypnotic state!

This explanation is unacceptable, because it poses more problems than it actually solves. The theory of telepathic information transfer between myself and the bookseller is perhaps stranger even than non-causality, but most of all, it does not explain why it should be this book, *Talking with Angels*. She had probably just received it, before my information started floating "in the air", taking into account delivery times. Unless my "information" travelled back in time? No, because then we fall back into non-causality.

The same problem is posed by the other coincidences: for example, we have to admit that St Teresa's quotation must have been in the church for a very long time, and that if its presence had had to depend on my mental state, then this state would have had to exert an influence going back in time!

Quite obviously, I reject such a theory of a piece of information being able to circulate in the air and travel back in time, even if certain faster-than-light theories in physics do allow for that 5. It is an explanation without really being one, because, once again, it poses more problems than it actually solves.

Now let us look at our most serious causal explanation, which brings into play an extrapolation of the observer paradox that involves considering that outside of our observations, the universe is not real.

The bookseller who replaces *Talking with Angels* on one of the shelves just as I am looking for such a book, after asking the "Angel" to place me on the right path, is likely to do this at any moment, without this changing anything at all in her life. Her life is not determined by such an action. Even if she does it consciously, the moment she chooses to put the book on the shelf will have no influence at all on the course of her own life. Consequently, this moment is in some way "free". Similarly, the title is of no importance to her: whether she puts *Talking with Angels* or *Prince of New York* on the shelf at the precise moment I am standing in front of it has no influence over her existence. She habitually puts away any book several times on the shelves of her bookstore—that is her job—and it does not change anything in her life, a point I insist on. The same thing goes for her supplier: the fact of sending some box of books at one moment or another changes nothing in his life. He does it every day. Consequently, the title of the book sent on such and such a day to such and such a bookstore is free: it can be completely indifferent to the course of life of those through whose hands it passes, thus altering nothing among all the observations of the universe, thus altering nothing in the universe itself.

With a single exception: I am the only person in the universe for whom the act of placing a particular book in a particular place at a particular time will have some influence. This is so very true that it is even highly probable that the book you are now reading would not even exist had I not experienced that coincidence.

For the bookseller, the act of placing *Talking with Angels* on the shelf just as I was standing next to it was probably not something she experienced as an observation likely to have the least influence on her own life. Perhaps she would remember her gesture on reading this book? If the answer is no, then even if her observation had been conscious, it was not a case for her of experiencing or observing something. So what was it?

My answer is this: for her, it is a case of a non-observed fact, something that is consequently still likely to exist according to an infinite number of versions: for example, changing the title of the book or the moment she placed it on the shelf, etc., with all such variations making no difference whatever to the bookseller's life, to her progress along her own tree of life.

As we have already seen it, we can admit that there exists in our future an infinite number of potential versions of our existence, versions that might only vary by some tiny parameter of distance or detail for example, thus causing a tiny shift along another dimension.

We are encroaching a little on our theory here, but we are doing so while still respecting causality, that is to say without actually supposing that our intention has anything to do with it! I would even be so bold as to go further and say that, among all the potential versions of our existence, most are not particularly likely to disturb our everyday existence by modifying our life or anyone else's, with such a simple variation of a tiny parameter of distance or detail.

Let us get back to our bookseller, but this time from the point of view of my own experience. I am going to ask the following question: since no preferred version really exists of the bookseller's behaviour as to exactly when she is going to put the book on the shelf, and since all possible versions of this act really do exist because they have not been observed, how does indeterminist chance make us experience one more significant version rather than another?

The answer is as follows: so long as there has been no observation or experience, there is no chance, since all eventualities remain in an indeterminate state. So chance has no usefulness. So long as there is no one to observe that something in particular has happened, everything that might exist according to the laws of physics exists simultaneously and chance has no place in it.

Consequently, we have to deduce from this that the bookseller's action, placing on the shelves at precisely the right moment precisely the book that holds the most significance for me, does not depend in any way at all either on herself or on the supplier. It is in no way determined by the experience of these two people. For, as we have said, the exact moment the bookseller carries out this action is not activated as long as it is not observed. And so it is free to depend on any other experience, mine as it happens, for I am the only one to observe it.

Thus we come to recognize a paradoxical but extremely important result of modern physics: everything that is not observed remains in an undetermined state (wave function state). Only observed events exist in a determined state, their determination being triggered by the observer!

Careful though: this still doesn't explain the bookseller's synchronized action!

From the moment that I am actually the trigger of my observation, regarding this synchronized action, there is at first glance nothing to oppose the fact that I am the one who provoked this synchronization by means of my mental attitude! The problem is that this still doesn't explain the "how" of the synchronization. Let us note the following, however: if the bookseller's action had not been synchronized, then I would never have been led to make that observation!

So it would seem that it is my observation that triggers activation of a reality which, without it, would never have been observed! Careful, though, there is no intervention of intention here, where we maintain our immutable principle of causality! So instead of intention, here we will invoke the possibility that the observer is himself, in the present time, responsible for the final choice!

Which would be tantamount to attributing the observer with an ability to create reality!

Not only would that outstrip the modest pretensions of our Theory of Time regarding the observer's abilities, but what is more, it would contradict the most recent results in quantum mechanics, which show us, using decoherence theory, that the observer does not determine the content of what he observes. All he does is exert a choice and make it enter reality, knowing that it remains random, at least in the absence of hidden variables.

To conclude, what is missing is an element that could rescue causality, and that element is this: something inside me, in my role as an observer, seems to inform the universe that out of all the possible events that might happen at the moment I am observing, it has to choose the one that corresponds to what I am expecting!

Causality does not provide this element of explanation!

Whatever the case may be, do you really believe things happen like that?

Could the universe really be subject to our whims to such an extent?

This, then, is the paradox to which the logic of causality brings us: forcing us to recognize, by respecting implacable logic, the fact that we may have extraordinary powers, magical powers, that consist of creating the reality we observe!

And so causality ends up confronting us with the irrational.

Let us now see how we can interpret these nine coincidences in connexion with double causality. In order to do so, I must reexamine the context in which my intention to write this book was formed.

My initial idea was to write a book about synchronicity. I intuitively sensed that there was a mechanism there and I knew from past experience that writing them down would make my ideas clearer.

However, I was aware that I was sorely lacking in personal examples. I had already experienced such phenomena, but without ever making a note of them, and my memories were not enough to assemble enough material to make my experience credible. Which meant that my book would be founded on no more than incidents related by other authors. I did not think this was enough for me to get down to work.

Reading *Talking with Angels* while on holiday with Danièle had focused my ideas, and the love between us made me sure of her complicity, giving me the impulse to set about this process of experimentation, a process I presented to her as a sort of game.

If my intention to write was real, then by doing this I had rendered it impossible to realize within a causal framework! Because the probability of my writing the book without having experienced those synchronicities was very faint indeed.

On the contrary, the probability of my writing it as a result of such experience, implying the intervention of indeterminist chance, was high. And it was reinforced by the fact that upon my return from holiday, confronted by my children's absence, I would be disappointed and find myself with free time on my hands. I didn't know it, but we may say that the universe knew it, and that my future was already stored in its memory.

The probability was also strengthened by the fact that I was aware of all the conditions required for making the phenomenon appear: risk-taking, letting-go, unpredictable travels, journeys...

So there I was looking at a situation where the greatly increased chances of reaching my goal by purest luck were turning out to outweigh the greatly diminished chances of reaching it by some "reasonable"—or causal—method.

From that point on, the mechanism of all the synchronicities was itself simply a matter of course!

Several trajectories of life formed in the universe, originating from the most probable moment in the future that I would start to write, in other words, the following week, and converging onto different moments in the preceding week.

I found the tiny church with the annotated picture of St Teresa because the path leading to that village and then to that church, both places on my Tree of Life, saw its probability greatly increased compared to all the other possible paths that chance might have guided me to.

The bookseller was to place *Talking with Angels* on the shelf at a precise moment during her day, after taking delivery of several books. One may suppose that on her side, there was a window of about fifteen minutes or so. On my side, I might get to Digne and walk into a bookstore, as I often did, at a much vaguer moment during the day, let us say with a window of about three hours or so, overlapping with hers. This meant that there was a probability of about 1/10,000 that her action would be synchronized with mine.

However, since it was an event likely to make an impression on me and send me on my way in quasi-certainty towards realizing my intention, the probability of this convergent, synchronous trajectory of life coming back towards my present at this very moment was greatly increased.

But it did not even have to be anywhere approaching certainty, in fact. All it had to do was remain improbable, while outweighing the probability of the best competing scenario, whether causal or not.

If, for example, the probability of my writing a book without experiencing synchronicities was one per cent, then all it needed was to be increased by a factor of a hundred, not a thousand! Especially seeing as the following week it was, on the contrary, quite probable that I would start to write, with or without a crop of coincidences!

Of course, these figures are merely estimates. They have the advantage of materializing a schema for a very much simplified but rational sort of mechanism.

We could consider the two "literary" synchronicities that followed as being "hooked up" to the same convergent trajectory. My intention must have generated a large amount of "wellspring water" the following week, and this water had flowed back towards my present forming "puddles" in certain spots. In fact, it was simply a question of there being moments on this trajectory when chance came to reinforce my own interpretation that I had to write my book, if only I managed to reap a sufficient crop. It is astonishing, however, to see just how effective the Law of Convergence can be. Out of all the possible chances of my random paths, it chose one that brought together all the "pieces of chance" related to my initial intention, making them all linked together on the same convergent trajectory towards my past.

Just like a river sometimes forms water hollows!

Now let us return to the most moving synchronicity I experienced, the one I encountered at kilometre 191,191.

There was something more than mere synchronicity in this fabulous convergence. For everything happened just as if the Angel existed and was talking to me!

The gîte that corresponded to my expectations was indeed a synchronicity, just like the previous ones, my intention of writing here replaced by my intention to have a good time on holiday with my children, somewhere particularly agreeable.

The fact that the gîte was empty apart from the proprietors, however, seemed to be "trying" to bring me information from the future, about the absence of my children.

This sort of information had been absent from my "literary" encounters. When we deduced from them that "this meant I had to write this book", it was not an accurate interpretation, for no one was giving me any advice whatever in this case. Even the Angel himself, even assuming he did exist, would not have had such a role, as he explains in *Talking with Angels*:

"Humans retain their free will and their choices."

There had been a unique, residual strangeness to my adventure, then, concerning the absence of my children, something that double causality did not seem able to explain.

Unless the Law of Convergence does not bring information with it?

A sort of "Law of Love" rather, that carried information to reassure or warn me?

I thought this completely ridiculous!

One might point out that this is already the case for all of these synchronicities, because at the same time as they reveal an intention, they carry with them the information that a path exists for realizing that intention, while also inviting us to follow it.

However, when the Law of Convergence has for effect to carry other information than that relating to our intentions, then there are grounds for asking ourselves some serious questions. And yet, this is indeed the case.

Before broaching the solution to this enigma in the last part of the book, we will now illustrate an example of information being brought to the past, using a set of facts I experienced personally, and these are:

The recurring double 22, which, in contrast to the synchronicities I have just set out, actually bring us proof!

Chapter 14—Summary

It is irrational to maintain causal reasoning to try and justify an avalanche of synchronicities.

On the other hand, it is entirely rational to employ doubly causal logic, whose simplicity and logic are evident.

XV.

THE DOUBLE 22

*In which I reveal how I experienced a second set of events
which mean we must bury causality once and for all!*

If it really was the series of synchronicities I lived through on the Road of Time that gave me the impetus to write this book, then it was above all a number of numerical coincidences involving the number 22, and more particularly, whole series of number 22s, that brought me the proof I personally needed to get down to work. As a scientist, I would never actually have risked elaborating such an ambitious theory for explaining the mechanisms behind these phenomena, had I not first formed a deep-founded conviction due to my own experience. The proof had long existed in books, but it remained unexplained and stood as a challenge to science. The fact that this challenge was personally addressed to me "by the universe" using series of the number 22, like some sort of provocation that could not possibly fall on deaf ears, was what led me to reveal it using my theory of double causality.

Several years before starting to write this book, I took something of a risk in life by deciding to start a business to exploit the results of my research into artificial vision.

My business associate, someone well versed in numerology, commented that the date it was set up, December 22, was the same number as my personal "path of life" as well as that of the business itself, numbers calculated using the letters of our respective names. So I started to look on this number 22 as a sort of fetish, and began using it all over the place whenever I had to choose a password or play the lottery, for example. I even went so far as to do 22 press-ups in the morning instead of 20, although it did stop there!

At this point, it wasn't yet a symptom of mental deviation, because there was nothing superstitious about it, just my passion for numbers and numerical coincidences. These had always interested me, and even though I hadn't ever actually experienced any myself, some sort of intuition made me think that by "sowing" this number 22, I might harvest something,

169

without too much conviction but somewhat in the spirit of "You never know if you don't try". I was still a long way from imagining that I could manage to provoke coincidences like this.

So I was utterly astonished when, a few years later, an article appeared in the local paper on the 02/02/2002 about my research, referring to my business, with the dateline above it showing this series of four 2. This could be a mere coincidence due to "chance", but the fact that it happened after I had "sown" the 22 was enough to awaken my suspicions. All the more so since it was the first time that my business, founded on a date containing a double 22, had been made the object of an article in the press.

All of this could have been quickly forgotten, but the same year, while several upheavals were making changes in my life, I noticed that the most important days, and the ones most emotionally charged, were all punctuated by some sort of observation of the number 22. On such days I would be literally pursued by this number. I often parked opposite the number 22 on a street. When I looked at the time, it would be 11h22 or 12h44. I would get up in the night at 1h22 or 2h44. I would open a book and happen on the page 22 or 122. I would pass a motorway sign showing 22 or 44 in the minutes on the time display. 44 often appeared as an alternative to 22 (11 or 88 were much rarer). And lastly, the person who contributed to a change in my life that year was born on April 22.

As a scientist, I was quite objective in analysing the fact that I myself might have a tendency to project this 22 onto my observations, that is, favour it without actually realizing it. So I remained dubious, but this interpretation was not confirmed by what followed.

2005, another year of great change for me, was marked by encounters with two people who became friends after offering me a great deal of support. These two people were born, one on September 22, the other on November 22. It occurred to me to check the date of birth of a friend who had helped me more than anyone else in the past, and I found that he had been born on January 22!

Now, this doesn't mean that one has to be born on the 22nd of the month to be one of my friends! Because there is still the question of helping me change my life, and to be perfectly clear, of the four people who contributed most to doing so, all born on the 22nd, the help given by one of them, I won't say who, turned out to be rather despicable in nature.

The following year saw the beginning of a real liberation and more particularly a positive evolution that allowed me to progress on a personal level to a point where I could realize my project for writing this book and settling down in Haute-Provence. One person accompanied me throughout this journey, just like a "lucky charm", sharing the experiences I describe in this book. Danièle was not born on the 22nd, but it is quite remarkable that our encounter should have been accompanied by a double 22 marking the time and date the first time we met.

During this transitional period, a double 22 I have already mentioned (D22 + 22 euros) was served up to me on the Road of Time, the very day when, finding myself assailed by all these synchronicities, I decided to write this book. I started to realize that this double 22, or the series of four 2, must have some significance. But who sent it? I was well aware that they came about at important moments in my life, when I had life choices to make. And I admit that for a certain time, the idea that there might a "little angel" sending me signals did cross my mind.

This interpretation came to me for the first time when one evening I returned from a business trip to San Francisco during which I had had to make the "heroic" but depressing decision to refuse to sign a huge software contract which would have had a lasting impact on my business activity, only to find that I had been given the number 22 at the airport bar for my turn to be served and then the number 22A as my seat number on checking in. I took it as a sort of consolatory signal meaning "Be happy, you made the right decision".

Another time, when I was moving into an apartment in Marseilles and was busy shopping for food to cook myself dinner in the new place for the first time, my supermarket bill came to 22 euros and 22 cents, like a greeting welcoming me to my new home.

Up until then, I still hadn't understood why I kept coming across these double 22s every time an important change happened in my life, and I still had my doubts, but I was starting to jot them down with a view to including them in this book. I had found an explanation for synchronicities, but these series of 22s resisted my attempts at analysis, and titillated me in a provocative sort of way.

It took a subsequent encounter with a quadruple 22, on the symbolic day that I committed an act of free will that remains perhaps the most important in my life, for me to receive the "shock" that led me to "reshape" my budding theory and activate the definitive writing of this book. I started finally to understand that it was necessary to draw a fundamental distinction between coincidences and synchronicities.

This quadruple 22 happened to me when I went to see a solicitor to sign the final act of purchase for a piece of land on the geological reserve that I had already spent a long time trying to find, and that I eventually did, in accordance with my own strict conditions, and by sheer luck.

I have modified the description of the geological data of this quadruple 22 somewhat, but it is entirely equivalent to the original.

That day, I was leaving the property that I had just acquired and turning the counter on the car back to zero to see how far it was from Digne, allowing me to see that it was 2.2 km from the main road, although the fact made only a brief impression on me.

Long before I got to Digne, and at the very moment the clock showed 22.0 km, I came to the crossroads with the RD22! So this departmental road n°22, which already stood out because of its "supporting role" in my decision to write this book, was situated exactly 22 km from my future home! I was happy about this double 22, and resolved to include it in my book, because it had declared itself on the very day I was once more set to change my life. But I was so used to this sort of thing by now that I told myself, "Could do better!"

What I was waiting for was a fourth occurrence, for I was unsure what to do with that initial "2.2", wondering whether I ought to consider it an authentic 22, or discount it altogether! I had never experienced a triple 22 and would have liked there to be a companion for the isolated 2.2. For a quadruple 22 would indeed have been worthy of the importance of the day's event! In all seriousness, I had had enough of the "little angel's" obscure wee jokes, and needed more than that to get me going!

Having parked opposite the place my meeting was to take place, I had fifteen minutes to spare and wandered through the centre of Digne. I entered a bookstore to buy some pens. As I came out, I noticed Bernard Werber's latest book, and checked the dust jacket to read the blurb. No blurb, but I noticed the price: 22.90 euros. With an inner smile of satisfaction, I told myself I had found my fourth 22, but I was not really content, because nothing really connected this 22 and my original 2.2. However, checking the book's publication date, I opened the back cover page, and chanced on the publisher's address—22 avenue de Huygens!

So that day, I was given a quadruple 22 made up of two authentic double 22s, that is, associated one with the other.

It was then that I started to think that my coincidences concerning the number 22 deserved their own chapter, but I knew I had to do some serious work to elucidate their mechanism, otherwise the theory I was trying to construct would not stand up.

Prior to the episode with the quadruple 22 that finally convinced me that I had to return to all my aborted efforts, I had on numerous occasions acted like a true sceptic and tried to reject the number 22, starting to find it troublesome and convinced that I was actually projecting it all over the place myself, by being over-sensitive to its occurrence on the important days of my life. This explanation, however, did not stand up to the calculation of probabilities, even conditional ones, and a true sceptic must know how to look at things both ways. This quadruple 22 reduced my scepticism to zero. All that was left for me to do was find the proper explanation, and the necessity of doing so stood before me like a veritable intellectual challenge that I simply had to take up.

Not for one minute did I suspect that after writing several pages on these series of 22s, the most surprising was yet to come. First of all, I remarked that there was a double 22 lurking in the satellite coordinates of my new dwelling, whose geographical position, at 44° latitude, 2.2 km from the main road, 22 km along the RD22, turned out to contain a total of six 22s, two of which were authentic double 22s! Added to this was the fact that the house is 22 metres wide and contains within its grounds one swimming pool and two springs, that the swimming pool is 22 metres from the house and the nearest spring 220 m away... The most edifying occurrence of all, however, is that the second spring is a genuine geothermal hot spring, flowing at a constant year-round temperature of 22°!

And that is not all: inside the house, all the radiators display a double 22, the first one engraved on the screw at the connecting pipe, and the second inscribed on the thermostat! Finally, the main living room is vault-shaped, 4.4 metres wide and 2.2 metres high, which, it has to be said, is not exactly your standard sort of height.

All in all, I uncovered no fewer than fourteen 22s in relation to the house, of which at least four were authentic, double 22s, grouped into sets of two 22s.

This time, then, it was quite impossible that I myself had generated such an avalanche of 22s, because this final series was, quite literally, set in stone! Something was resisting my analysis!

Despite all the conclusions these observations were leading me towards, I still had trouble understanding the role my intention might be playing in all this, and as soon as I started looking for some mechanism independent of my intentions and projections, I was lost!

I found inspiration for the solution to this problem via a friend, an astrologer and numerologist, to whom I had recounted all these phenomena and who said to me one day:

"It's hardly surprising that you find all these 22s, because 22 is your path of life!"

He was talking about my numerological path of life, something I already knew, but I couldn't see how merely being aware of this path of life could generate all these 22s. The light started to dawn when he added:

"What's more, the longitude of your natal sun is 22.44°."

So there was already a double 22 stamped on my birth chart concerning the position of the sun, and this was something I had been completely unaware of!

Yet this wasn't all, because when I remembered that there was another 22 in my time of birth, which was 22h35 after calculating the time difference, it then occurred to me to calculate my solar birth time, taking into account the time difference between the Paris and Greenwich meridian lines, which happens to be 9 minutes.

Thus I discovered that at the time of my birth, not only was the sun at 22.44° of longitude on the ecliptic, but also I was born at 22h44, solar time!

At my birth, then, there had already been a quadruple 22 characterising the position of the sun at that precise moment!

This observation was a "eureka" moment for me, rapidly leading me to understand what the mechanism was, for it began to be clear that my intention alone could no longer, in any non-causal way, explain the link between the 22s of my past and those of my future.

It was no longer possible for me to think that, because I was "branded" by the number 22, I could intentionally have generated all the other 22s, through the intermediary of the same mechanisms as synchronicities or, worse still, through projection!

For I had been ignorant of most of these facts about my birth and this could not in any possible way explain the astonishing concentration of double 22s relating to my new dwelling.

Indeed, it was quite impossible to imagine that the six double 22s concerning my new home might have been intentionally provoked, because they were related to a house more than a century old, that had already existed before I was born.

So, in order to explain such a highly concentrated order, the Law of Converging Parts was obliged to propagate it into the past! The eventuality had to be envisaged that the two double 22s concerning my birth might themselves be caused by the high concentration of 22s observed in my future. For in fact it was much easier to imagine that there were sufficient variables present at my birth for the quadruple 22 to have come from my future. Shifting my birth a few minutes either way changed nothing whatsoever for the universe and its observers. The indeterminism of my birth therefore could perfectly easily be completed by a piece of information travelling back in time, exactly as is the case with synchronicities.

Except that, in the case of synchronicities, observation always came after the intention that acted as a catalyst for it, meaning that some significance could be attributed to it! Now, I had already understood that a coincidence could differ from a synchronicity, in that observation of coincidence could precede intention, thus preventing the observer from understanding its significance. But I hadn't yet understood that by travelling back in time, the Law of Converging Parts could complete indeterminism even in the past itself, by creating meaningful order!

For it seemed to me that it was impossible to modify the past! And yet this double 22 surrounding my birth suggests quite the opposite, while at the same time giving us an explanation: it is entirely possible to modify the past so long as such modifications concern aspects of the past that have never been observed, as though they didn't yet exist!

And this is how, just as with the future, we observe that part of the past can very well remain completely undetermined!

So time must be much more symmetrical than we think!

Therefore, the past might be modified, but only, I repeat, on condition that traces of it had never as yet been observed! The "modification" of the past would be a result of the absence of sufficient traces to allow its definition, absence or even actual disappearance, in fact. Why not imagine that the Law of Converging Parts, the law that creates order, might itself be able to define what will be observed when the past is concerned, since this is already the case for the future?

However, let us come back to the mechanism for making these double 22s appear, because I hadn't finished with my explanation.

What characterizes these double 22s and my house are material facts, fixed and independent of any vagaries, especially of time: satellite coordinates are impossible to modify, as are distances and size! Give or take a few decades, these numbers stay the same, because quite purely and simply, they characterize this house and its grounds.

On the other hand, it is quite another matter for the other double 22s that preceded this final accumulation. A day or even a few minutes more or less either way, and they might well have disappeared altogether. One only has to review them to see this, starting with the one concerning my birth.

It is clear, then, that if there is a link between all these double 22s, then the information that created this link travelled from the future towards the past, and was not content to stop when it got to the present.

The past was rearranged in such a way as to transmit the following information: a change of life concerning me, realized through my own free will, is accompanied by observations of double 22s!

Here at last, then, is how I explain how it happens:

On the occasion of the conscious double 22, when I created my company (and then began "sowing" the number 22), my as yet vague intention to "sow" this number was able to start "feeding" certain parts of future space-time that were compatible with the information and that I might encounter along my route.

I couldn't say whether the first "echoes" I received came as a result of this fixation with the number 22, but in any case, it only required a few true or false "echoes" during 2002, a year more propitious, it has to be noted, for such apparitions, for me to increase the significance I was then beginning to attribute to series of 22s, in connexion with the changes in my life that started that same year. Being intentional changes, this time it was quite clear that they acquired the ability to feed the "sources" of my potential futures that might be compatible with this double 22. I had then been led to encounter each of these sources, leading right to the main source of all these double 22s. The closer I got, that is to say, the more I increased the probability of encountering the source one day, the more the source itself had a tendency to make me encounter others leading back to it, and this is why my observations grew stronger and stronger as I gradually grew nearer to discovering it.

What was astonishing about this source was that it symbolized a concentration of double 22s that had been carved into stone for many years, and on top of this there was an actual source, at a temperature of 22°!

What we have here is nothing other than a retroactive loop being gradually set up between my future and my present, at certain moments of the present when I received echoes from the future!

In a similar way as with synchronicities, this retroactive loop resulted in generating a phenomenon of attraction between the trajectory of my life and the potential trajectories that might be sources of double 22s! Meaning that I did indeed reap what I had sown!

I settled on the following conclusion, the most logical: I had quite literally been "drawn" towards sources of double 22s, and the phenomenon had been powerful enough to be responsible for the chance that allowed me to find a new house, just where they were probably concentrated at their highest level!

We will subsequently see that this phenomenon of "attraction" towards situations that reflect our thoughts is actually quite a general one.

Chapter 15—Summary

1. The author has experienced successive, powerful coincidences in the shape of series of "double 22s", of which the most recent were observations concerning his past, something tending to the conclusion that the "double 22" information travelled back in time, from the future towards the past.

2. The possibility of modifying the past: in the absence of traces of the past, the past can be modified by the Law of Converging Parts according to a process for maximizing order that can explain the "double 22s" that travel back in time. **The result of such a mechanism would be to recreate the past by gradually simplifying it, as and when its traces disappear.**

PART IV

REVELATIONS

XVI.

INNER SPACE

In which we discover how a theory of physics quite naturally invites us to become free men.

In the first two parts of this book, we have dealt with double causality by presenting it as a Theory of Time with the help of metaphors about the Tree of Life, a network of rivers and the Road of Time.

However, all these metaphors are essentially spatial ones. So now it is important to justify them by introducing real, multidimensional space.

The reason we needed these metaphors was because of how difficult it is to represent such space, something our theory requires in order fully to describe the extent of our potential lives. For we supposed that all these lives were not merely possible, but that they already existed in present time, and thus already interacted with our present-time consciousness, able to modify their probabilities of being actualized.

Without bringing space into it in a more rigorous manner, our description of the mechanism of the influence of intention over the chance of non-causal opportunities had until now been limited to a pattern of temporal bifurcation. This pattern can be summed up as follows, by bringing in two factors of bifurcation:

– the activating of a new future by sufficiently increasing its probability of being actualized;

– the deactivating of an old future by diminishing or even cancelling its probability of being actualized (point of no return).

When the probability of a new potential future can compete with that of the future which is already activated, then under the influence of our free will and our observations the balance may change, resulting in our being directed towards a new trajectory on our Tree of Life. This, at least, is our present hypothesis.

This new way of presenting things implies that most of the time, we are moving along on a track, half-asleep, and that it isn't easy to exercise our free will by intentionally activating a change of direction.

Such a mechanism remains simplistic and inadequate, however, because we have omitted to describe the spatial part of the processes which come into play, and this is what we are going to set about doing now.

In order to arrive at a unified theory of space-time, physics has produced several remarkable multidimensional theories, including that of Kaluza-Klein in five dimensions and String or M-Theory in 11 dimensions, which benefit today from the new lighting of the Maldacena's conjecture. What lies at the origin of physicists' need to introduce extra dimensions to our four-dimensional space-time is the famous indeterminism of quantum mechanics, which is interpreted by our own work as a limit of density of information into the universe[6].

The most important consequence of this sort of theory about our conception of space is that in order to interpret these invisible, extra dimensions, it leads us to envisage the existence of "parallel universes". These universes will somehow contain all the other possibilities of evolution that our own universe might have had, given its indeterminism.

In order to represent all these universes within one space, String Theory calls on six extra spatial dimensions, folded in on each other and invisible because they are so extremely small. Everything happens as if every line in space behaved like some excessively fine cable: meaning that we can move around this cable an infinite amount without changing our position on the line. If we then roll each point of the cable in on itself by decomposing it into an infinity of small, closed strings, we obtain two extra dimensions for each line. Given that our space is already made up of three dimensions or lines, this brings its number of spatial dimensions to nine.

From a mathematical point of view, we can thus attribute a certain number of extra, invisible dimensions to each point of physical space, representing as many possible versions of what is happening at that point, which is why these extra dimensions lead us to conceive of parallel universes.

This is made clearer with another image representing the same theory, consisting of acting as if we lived in two instead of three dimensions, in other words, on a flat surface labelled a membrane. Our parallel universes would thus be all the other membranes parallel to our own, and invisible to us, all fitting into three-dimensional space.

What we will now do is suggest a more concrete, and above all more realist, representation of these parallel universes. Let us return to the

---

6. Ph. Guillemant et al, *Characterizing the Transition From Classical to Quantum as an Irreversible Loss of Physical Information*, arXiv:1311.5349 [quant-ph], 2013.

multitude of potential lives deployed on our Tree of Life. If, at every potential bifurcation along this tree, we have a choice between two lives, then we will find ourselves with a thousand potential lives from the time of our birth. But if we want to memorize all these lives in a Cartesian space, we have no choice but to call on parallel universes: one universe for each life. From this point of view, one universe thus contains a possible version of all our lives as a whole. In this way we can understand why, taking into account the unimaginable number of all the possible versions, we have to add dimensions to space.

Coming back to the subject of our free will and the link it may have to this space, we will now show that the use of intention that generates the shift between two life trajectories has one effect, that of quite simply making us "change universes"!

This may seem incredible to anyone imagining that one would have to take a spaceship and visit the edges of the universe to change space. But not at all, it's much simpler than that.

First of all, there is a very simple and rational way to begin to accept this change of universe: and that is to consider that all these universes are so parallel to each other that they are practically identical to our own!

In which case, the shift in the universe obviously has the same effect on us as staying in the same one: i.e. no change.

But in order for this possibility of really changing universes to correspond to a bifurcation on our Tree of Life, we have to consider the existence of at least two distinct parallel universes, associated with this bifurcation: the first, in which I realize my old destiny, and the second where I realize my new one.

Now we can understand how our Tree of Life, with all the possibilities of different existences it potentializes, is only describing all the possibilities of parallel universes that we may visit during our lifetime.

Such an interpretation of parallel universes in the form of a collection of trees of life offers the immense advantage of finally lending an interesting meaning to this theory for us as humans. For its main drawback had been in not giving us the correct reason for all this wasted space. Now we can understand that this correct reason is actually our own free will.

Now that we know that, out of all these universes, there is one privileged universe that hosts our present lives, we can better understand why it is important for the others to continue to exist, insofar as we may well shift universe at any moment, all together. And I did say all together, for when at a given moment you alone change your life, you take everyone else with you into another universe.

Let us just open a parenthesis onto one consequence of this vision of things, one that would take too long to investigate: the fact that changing universe will also modify all the potentials that were associated with it, for the reference point from which these potentials derive will become our new universe. So it isn't simply a question of changing place, for we are not certain of being able to come back again, something that would suppose that the old universe would once more be part of the potentials. Changing universes is a serious business, then, for new universes will appear that didn't exist before, co-created by our new possibilities, and all the universes whose existence depended on the old one will disappear: we are watching a genuinely irreversible process!

So this is where the true irreversibility of the universe dwells: our simple position as an observer, deciding to take one bifurcation rather than another, generates a process of co-creation, which finally explains how our thoughts create the world, even though all we are doing is observing!

Supposing that this immense space, infinitely multiplied, created by this infinity of parallel universes in a state of permanent co-creation, is in the end no more than a space dedicated to our trees of life, we can now understand why there is actually no wasted space, just an immense, dynamic variety of potential lives. However, all these lives have to be memorized somewhere. It isn't us who need this, but the universe itself, for it seems unrealistic to prevent it from existing after such an arbitrary moment as the present.

In fact, what other alternative do we have to parallel universes? Surely it is preferable to get used to the existence of these multiple universes, hardly such an unacceptable concept after all, rather than suppose that not only our entire universe, but also an infinity of universes have to be recreated every second by the passage of time? Luckily, we have already seen that this is not the way things happen, and this, by the way, was Einstein's main preoccupation.

For Einstein himself said: "The distinction between past, present and future is just a stubbornly persistent illusion. Time is not at all what it appears to be. It doesn't flow in just one direction, and past and future are simultaneous."

With this new vision of things, which merely borrows Einstein's conclusions, but in a space multiplied as many times as necessary to allow our free will to express itself, we will now be able to establish a much better idea of the mechanism of the intention that causes us to shift to another life, or, in other words, to move from one universe to another.

Remember the fundamental assumption of our Theory of Time: the reality of our free will. It is this free will which, in the presence of potential bifurcations in our future, forces us to consider that we do not have one life, but rather an infinity of potential lives, of which one is already "materialized" in our present universe: this is the reference point from which all the other lives derive.

We have therefore reformulated our assumption into an ability we possess to move, at the same time as in dimensions of our physical universe, within the extra, hidden dimensions of all our parallel universes.

Let us take stock of what we know right now:

We already know that our body allows us to move in three physical dimensions.

We also know that the other dimensions of space are invisible, because they are folded in on themselves or excessively small.

We know that our free will allows us to move into these extra dimensions.

We know that we move in this way by means of our intentions and observations, controlled by our free will.

We know that our intentions and observations are part of our conscience, in other words, part of our inner space*.

We can deduce from this that the extra dimensions of space are none other than inner dimensions*!

Which means that there must be a part of us that can reach into these dimensions, one that by necessity must be timeless, because in order to place an intention in the future without interacting with the present—something which would require a forbidden exchange of particles—part of us must already be there, that part, as it happens, that embodies all the probabilities our intentions can modify.

And it so happens that we have already labelled this part of ourselves in our model of the Spirit: we are talking about the Angel itself.

Thus the minimalist version of our theory does very well in describing the Spirit we address in *Talking with Angels*, a spirit which turns out to be nothing other than a supra-dimensional part of ourselves. No need for any notions of incarnation!

We will now break new ground in order to reach a second fundamental consequence of our assumption about free will.

Remember that the first consequence of this assumption could be seen in a modification of our concept of time that led us towards double causality.

The second consequence of this assumption will now affect our concept of space, and lead us to the notion of **inner space**.

In physics, the introduction of such an inner space is long overdue!

However, as this is an attempt in which the formal details and consequences go largely over my head, all I will do here is sketch out a few elementary principles.

One of the first principles consists of representing our intentions within the space of our inner dimensions, using information located inside this space, some sort of fluid, energy or matter, whatever it may be, but something that above all memorizes those intentions.

As our intentions, by their very "definition", contain the things in life we want to see realized, then "by definition" of our Tree of Life itself there has to exist some representation adapted to the "materialization" of these intentions: this translates into one or more places, parts or even zones of the space of the Tree, corresponding to the things about our lives that we "project" onto them.

We should not find this notion of materialization shocking, because modern physics has already made us used to the idea that the matter of our universe has no "solid" existence. Indeed, we are noticing more and more that **matter is actually nothing more than information**, and the "solid" aspect of it comes only from its interactions.

By materializing our intentions in zones of our future potential spaces, we simply express the fact that we are capable of watering our Tree of Life in these zones, by sort of moving our "watering-can" over to them. The location of our intentions in these "watered" parts of space in our parallel universes poses no problem as long as it is simply a question of probabilities of potentials, in the same way as the location of a particle in our physical universe poses no problem as long as it exists only in the form of a distribution of probabilities of presence, whatever the moment may be. In this way, we express the fact that as long as we are working with probabilities, nothing makes it obligatory for the things we are describing to be happening in the present. If I have the intention of buying a book, the probability of my buying it does not increase at the moment I make my purchase. It actually increases at the moment I form the intention of doing so.

When we described the mechanism that lets us memorize these intentions in the future, we effectively employed the notion of probability, the programming of our intentions consisting in fact of increasing the probability of potentials occurring in a zone of our Tree of Life.

It is entirely rational now to admit that, supposing our free will is real, we have at our disposal the means for "materializing" this free will through an instantaneous increase of the probabilities associated with one or more zones of our inner space. All the more so since our brain is capable of reprinting any such memorization it keeps in its own memory at any given moment. During this process, however, the memorization is not only reprinted in our brain, but also in that zone of our inner space that belongs to our future.

This is the point at which new, unknown possibilities of our "brain" can be distinguished! They are only "apparent" properties, for they belong to our inner space.

If you were to ask a physicist or a mathematician to make a model of this process, he would call on a digital simulation that would consist of filling a region of our inner space with a series or chart of numbers, and if you then asked him to visualize that series, he would make it appear in a more or less coloured form, the colour corresponding to the density of probabilities.

Exactly the same way we would represent a form of fluid energy in our physical space, a cloud in the sky, a layer of water on the earth, or a layer of oil underground.

So what purpose could these extra dimensions of space serve, if not to memorize this inner information, just like our physical space memorizes matter? All dimensions are only useful for one thing, to store information in the shape of matter, matter which, as we have seen, can be entirely reduced to information.

And our probabilities, too, are precisely that, mere information.

If we sum it all up, then from a physical, mathematical point of view it is perfectly legitimate to lay out the following principle:

By increasing the probabilities for future occurrences with which they are associated, our freely chosen and clarified intentions translate into the displacement or the appearance of fluid layers in certain zones of our inner space. Let me remind you that these zones are not "'situated" in our brains, but somewhere in certain space-times that are invisible and parallel to our present space-time.

Which means we are quite literally capable of making it rain in our future! It is no longer even a question of metaphor, for our intention has actually planted, displaced or shaped information in one or more zones of our inner space!

Now let us take an example to allow us to get a glimpse of the fabulous consequences of this new observation.

If I have the intention of buying a book, all the existing future possibilities for me buying it will suddenly receive that sort of fluid from me — on condition that they are credible, that is that they fit into a scenario that gives me some probability, even an infinitely small one, of coming across this book. This means that a bookseller who happens to be somewhere I am likely to go, will also receive some fluid on his own Tree of Life, an increased probability of meeting me in his future, supposing that this holds the slightest interest for him.

Of course, this is merely an expression: the bookseller does not actually receive any fluid, only the time and place in his future space where we are likely to cross paths will receive that information, that "fluid".

But this does allow us to see that the fluid is already spreading to a Tree other than my own.

And that's not all. If possibilities really do exist that our paths might cross, then the scenarios necessary for these possibilities being realized, like the fact of simply going into town, will also see their probabilities of being actualized increased:

The fluid will start flowing towards the past!

And it is even possible that because of this, the probability that, while strolling through town, I might recall wishing to buy the book, is increased! What I mean by that is that we could ask ourselves if my intention needed to be memorized by my brain for that wish to come back to me!

Out of so many other probabilities, all the ones that involve my passing certain particular places where it is easier to find a bookstore will be increased : and so the fluid spreads further, right into the side-streets!

That, however, was not my intention, which was merely to buy a book, full stop! When I formed that intention, I was a long way from imagining that I would go into town, and turn down such and such a street in order to realize it. If for every intention we had to imagine a precise date and scenario for realizing it, then we would rapidly go quite mad! We wouldn't be able to work any more, because we would spend all our time imagining scenarios for all our intentions.

Which is pointless, because the universe does it for us, at least as long as our intentions are trustworthy!

So we have to admit that the fluid we have created in the future has started to flow back to the past all by itself, to the places we are likely to visit ourselves! So there is no point thinking about it, unless we wish to reinforce the same intention.

What, then, is this new piece of information circulating all by itself, without a by-your-leave? We are no longer talking about buying a book, but of visiting particular places that I don't even know about yet, so that the act of buying a book might be actualized. It is as if my inner space were working for me!

As it flows, the fluid may even encourage another encounter, an unexpected meeting, for example, like the one with a friend, a woman living in the part of town I would be walking through, and with whom I had lost contact.

The probability of meeting this friend is thus increased in a zone of our respective futures, without even her or me being aware of it, quite simply because by having this intention to buy the book, the fluid has flowed all by itself towards our possible meeting!

So the fluid began to flow, in several places, more and more abundantly, increasing the probability of scenarios I hadn't the slightest idea about!

Let us look back. Each time we form an intention, it now appears that it isn't necessary to think about all the possible scenarios for realizing it for them to receive fluid for their potential actualization.

Because our intention is trustworthy, all that is necessary is that the probabilities for realization that we have planted in the future are maintained coming back in time.

This is the moment we should point out that without the determinism of the Law of Converging Parts, limiting the number of solutions capable of reaching our present, these probabilities would become diluted and disperse. The importance of inverted determinism is thus to maintain the probabilities of realization when they come back in time.

The question we now have to ask is whether our intentions are truly active. Because for the moment, we have only talked about potentials, not their entry into reality.

So how does the process of actualization of possibilities actually take place, following all these modifications of probabilities?

Let us keep in mind the fact that all these parallel universes discovered by physicists are not a waste of space. If they exist to memorize all these potentials, then there has to exist some mechanism to allow one of these potentials to enter reality! Don't forget that their multiple existences are justified by the fact that the probabilities of these potentials may vary at any moment, according to the degree of precision, accuracy or even sincerity of our intentions!

In order to make one of these potentials enter reality, our free will still has a role to play. If it intervened at the moment we formed our intentions, then it still has to intervene at the moment we choose one path rather than another. Because the potential or potentials we have created will converge towards our present trajectory at different moments when possibilities for bifurcation are going to appear to us.

Out of all the paths whose probabilities have been increased through the circulation of fluid, we can take an uncontested guess that only the most probable will be detectable by our observation, that is, those that will have received the most fluid.

However, even if our mind is not receptive to the manifestation of these potential paths, it is practically certain that only the most probable causal path will retain our attention. We will have somehow asked, heaven will have helped us, but we will have done nothing to help ourselves.

Now let us suppose that several potentials appear to me while I am in a side-street looking for a bookstore: the possibility, for example, that I discover for the first time a little bookstore well worth getting to know better, or the possibility of bumping into my friend who lives in the neighbourhood, or even the possibility of my buying something other than a book, etc. I should remind you that these potentials exist without having been experienced, and that as long as I am receptive they subsist in a state of potential lives capable of being actualized, still worthy of receiving my fluid.

So while I am walking along, they may well come to my attention, if they capture my interest, by making me stray from the path I would have followed in the absence of fluid. So we are dealing with potential bifurcations.

How do they come to my attention? Let us consider a possible scenario.

After parking in the town centre, I walk along an avenue then cross at a pedestrian crossing to turn down a side road I imagine cuts through to the main shopping area.

Because there is fluid present from my intention to buy a book, barely fifteen seconds after I cross the avenue, my friend uses the same zebra crossing, but in the other direction. Had I been extremely observant, I might have seen her from afar. However, as the fluid is not very concentrated, I don't see her and she doesn't see me, because just as she might get me in her field of vision, I walk out of it on my journey down the side-street.

There is nothing surprising about this. To be sure of meeting up, I would have had to use the zebra crossing ten seconds earlier. A perfectly normal situation. Nothing different here from normal chance, in other words, something that accomplishes things fairly badly most of the time.

Now let us consider another scenario, in which we change nothing at all in the structure of the universe, except for one tiny thing: a few days earlier, my friend was thinking about me and telling herself how nice it would be to see me again. She wanted to ask me something, but because she had mislaid my number, she put off the job of digging it out. And yet, her intention did produce the desired effect.

For this simple intention began to set the fluid of probabilities flowing among all the possible scenarios for our meeting to be realized. It is easy to imagine that initially, these probabilities were tiny, but the day I formed the intention of buying a book in her part of town, the probability of the scenario of our meeting on that zebra crossing began to increase. And the day I actually walked through her neighbourhood, that probability suddenly increased, as if it were sucking up all the fluid that had spilled over onto the other scenarios.

And the closer we got to that zebra crossing, the more that probability increased, for the scenarios of chances that might yet have dispersed our trajectories were eliminated one after the other. This caused my friend to speed up, and me to slow down, by means of the smallest indeterminist variations on our respective paths, variations that might be due to nothing more than the presence of all the other pedestrians.

"Hi, Philippe!"

"Sophie! Well, well, what are you doing here?"

"What a cheek! I live here!"

"Er, yes, well, er..."

"Don't just stand there in the middle of the road, we'll get run over!"

"OK, come on then, let me buy you a coffee!"

"But of course! It would be a pleasure!"

Starting out with an infinitely small probability of our paths crossing, at the moment we planted our intentions, a probability which nonetheless constantly increased independently of any causal mechanism, my friend's initial intention, combined with my own, had quite simply ended up making our trajectories perfectly synchronized, giving us the impression that we had been drawn to each other!

Chapter 16—Summary

To develop a Theory of Everything, physicists have been led to attribute extra, invisible dimensions to our own four-dimensional space-time, and to envisage the existence of parallel universes.

Now, our hypotheses about free will and the omnipresence of the future also require the existence of parallel universes, whose purpose would be to host all our Trees of Life. Truly exercising our free will would thus lead each time to us changing space.

In order to establish this new concept wherein man plays a fundamental role in co-creating the universe, we are brought to consider that the extra dimensions of space are quite simply dimensions of our Inner Space.

The process of changing space-time by exercising our free will could be irreversible.

Put like this, our Inner Space would be the home of the circulating "fluids" that serve as a geographical location for our intentions. At every moment, our actions are capable of influencing this circulation of fluid, continually altering the "observable" potentials all along our path.

XVII.

THE LAW OF UNIVERSAL ATTRACTION

*In which we return to the finest law of physics and broaden
its field of application to include our inner space.*

Here, then, is how we can sum up the most unexpected result that can be deduced from the assumption of our free will, when we combine it, on the one hand, with the theory of parallel universes born of quantum mechanics, and, on the other hand, with Einstein's ideas on the illusory nature of the present and the simultaneous nature of the past and the future:

We can make it rain in our future!

Even though this rain is just a rain of probabilities, they in fact behave just like water flowing in space, except that they do so in the direction of the past.

In Part II, we already employed the metaphor of a source of water situated in the future, without even suspecting that it might find its fullest justification in the concept of inner space. This rain metaphor signifies that our new intentions "water" our inner space in places that correspond to the potential realization of these intentions.

The best example I have, the one at the origin of this book, was the focus of Chapter XIII: three years ago, my intention of writing a Theory of Time to explain synchronicities watered a branch of my Tree of Life at the corresponding times and places, most notably exactly where I happen to be writing at this very moment. This is what provoked the series of coincidences that I experienced at the time on the Road of Time, insofar as I let my decision to write depend on chance, and on chance alone. This chance replied to me like an echo being reflected back from my future, but with extra information as well! Isn't this a clever way of taking advantage of the properties of time to check the feasibility of our projects before getting started on them?

Taking this result, we will now reveal the most fundamental consequences of the Theory of Double Causality. They will cause the Law of Converging Parts to evolve towards a genuine law of universal, dynamic attraction*.

The reason it is necessary to make the Law of Converging Parts evolve towards a dynamic law is based on our observation that the probabilities associated with the different life trajectories we see before us are constantly varying in time and space. They vary right down to the finest details: not only do these probabilities evolve in time under the influence itself of our changing intentions, but when our intentions remain constant they still evolve according to our behaviour, and according to the intentions and behaviour of other people. And it is this continual evolution of probabilities that lends such accuracy to the metaphor of a flowing fluid, because the most powerful cause for these variations in potential probabilities is the "flowing" of time itself, as, by making certain potentials enter reality, it spends its time rendering certain scenarios credible, and eliminating others! So we shouldn't say, "time is passing!" Instead, we should say: "The chances of my intentions to be realized are passing!"

However, we must mention factors other than intentions that might intervene in this evolution, and that is behaviour. The very fact that we can never entirely master our intentions, because they are too often controlled by our thoughts, doubts and moods, means that we produce troublesome factors. They impose their own determinism and make it difficult to attain a state of serene wakefulness and relaxation conducive to helping us distinguish between our true intentions and our false, determinist intentions. What results is an unstable mental state that seems somehow to make the fluids spread, diluting them here and there and thus contributing to minimizing their effects.

This is why the positive, voluntary effects of intention are so difficult to provoke, unless one respects the proper attitude we summed up using several rules in our "Spirit Model". The attraction mechanism we will now address will foremost depend on having this proper attitude. Let us first consider the different conditions for its appearance, one at a time.

Firstly, we have insisted on the importance of letting go, the necessity of forgetting that intention so as to favour its concrete effects. At first glance, this might seem to contradict simple reasoning that would lead us to think that the more we kept our intentions foremost in our minds, for example, by imagining different scenarios for their realization, the more fluid we would produce!

In reality, it is quite the opposite!

Imagine that we were constantly to maintain the same intention, like a sort of obsession. Have you already tried maintaining the same intention over time? By asking this question, what I want to do is highlight the fact that in order to keep an intention in our minds, we cannot prevent ourselves from constantly modifying it, even slightly, even though it might be just at the level of scenarios for making it happen. If we were never to make the slightest modification to an intention, by always visualizing the same scene, for example, without ever looking into the least variant, then it would be completely nonsensical or even absolutely impossible to keep it in mind. When we transform an intention into an obsession, we cannot help but examine all the possible variants, the important thing for us being only the result, the goal to attain. And by doing this, we modify our strategy.

The problem is the same one that concerns an archer aiming at a target. If he doesn't let go, merely maintaining his intention to hit the centre of the target will constantly make him modify the position of his bow ever so slightly, according to the correctives endlessly dictated to him by his mind. This is not how top-level sportsmen and women work to achieve their goals and excite the passion of their supporters, a passion without which their supporters would never elevate them to "godlike beings".

It is clear, then, that obsessively, even merely consciously maintaining an intention can only have negative effects on the flow of fluid carrying the possibility of the right action or favourable chance.

The reason for this is that when we modify an intention, we cancel out the effects of the preceding intention in favour of the new one, so the fluid disappears from one place and spreads elsewhere—this being a fluid of probability and not a fluid whose volume is constantly being increased by being made to circulate everywhere.

For it is impossible to sustain the probability for realization of an intention at a sufficient level if it doesn't know what it wants because it is too busy spreading itself thin into the definition of its scenarios. When we think we are increasing our chances by trying to foresee multiple possible scenarios, in fact all we are doing is diminishing them. On the contrary, it is much more preferable to remain utterly imprecise about the scenarios and define only the goals themselves. For the scenarios are merely a means to an end, and defining such means only calls into question, once again, the non-causal realization of our intentions, by placing them under the control of mere causality. Thus, intention becomes slave to all the means we associate with it, insofar as it unconsciously incites us to favour these means at the cost of any other scenario. Goodbye chance, hello planning!

So it now appears that letting go is a guarantee for the new potentials revealed by our intention to be not only stable, but protected from ourselves. That way, they can automatically grow when the possibility arises for them to be further increased, just before they are actualized. And we cannot foresee that moment, because it depends on details of the opportune structures of the universe, and in no way on our plans. All we have to do is be ready, without forgetting that not only must our intention rest peacefully, but our behaviour must stay in harmony with the intention that has already been planted.

Apart from letting go and behaving so as to favour the emergence of opportunities, these being the two principal conditions, we have mentioned others, specifically: a genuine need for help, a real preoccupation, risk-taking, and new, original intentions.

The need for help and preoccupation reinforce the importance of associated intentions, because they take on a vital aspect. We might conjecture that they help to create a great deal more rain, increasing the probability of our realizing our objective. Risk-taking, as we have already seen, diminishes the probability of competing scenarios of a causal nature. It corresponds to a gradual abandoning of causal paths, without there actually being any alternative solution as yet. Having new or original, but always trustworthy, intentions, corresponds to a genuine, active change in the distribution of probabilities of potentials, likely to appear in the subject's path.

These last three cases, then, represent important conditions favourable to second causality, and we can spot a remarkable common denominator here: the emotional factor.

The need for help, preoccupation, risk-taking and innovation are in fact bearers of emotional instability, if only in that they make our future path in life much more random and therefore worrying: we don't know where we are going! **Emotion**, then, appears like a sort of "receptacle" that keeps **openness to change** in our life and is absolutely intent on keeping it that way for as long as we do not know exactly how we can realize the change!

Now we have already observed that this ability to remain open to change is an entirely positive thing! So emotion is excellent, precisely because we are dealing with a complementary factor to intention, especially when we want that intention to be realized with the help of "second causality", i.e. with God's help!

To sum up these two factors:

– intention increases the probabilities of all the scenarios or "branches of life" that can appear to us for realizing them, whether they be causal or non-causal;

– emotion "keeps intention" and all of its potential, when there is no causal scenario imaginable and we therefore have no means for realizing it as yet.

For we have to admit that the only reasonable attitude to adopt, when we have intentions that seem unrealizable, even with patience, usually consists in abstaining: but emotion stops us.

If we do not alter our intention, then we will find ourselves facing a situation that will necessarily maintain our emotion, because of its lack of reason and because of the contradiction inherent in maintaining an intention that is apparently contrary to everything that is feasible. On a mathematical level, with intention increasing the chances of non-causal solutions and emotion diminishing the chances of causal solutions, the combination of these two factors increases that much more the probability of the emergence of the non-causal path or, if one prefers, the path that depends on second causality, i.e. God's help.

Careful, though: as long as emotion merely keeps up the energy of the intention, we find ourselves dealing with an intention that remains in our mind, something which is not propitious to its effects actually being realized. Emotion must limit itself to a role of simply accumulating energy for the time needed for the intention to "resist" reason, or more exactly a "reasonable" attitude. The intention must in fact resist the temptation to be abandoned and it cannot do this without emotion. Depending on the case, this can lead to two forms of letting go:

– a letting go leading to a genuine abandonment of the intention's project, wherein its potential effects are cancelled, because we will no longer even pay attention to the opportunities for realizing it that might still follow;

– a letting go that comes from realizing one is temporarily powerless, accompanied not by abandonment, but by salutary hope!

This hope, despite the end of that emotional "rumination", has the advantage of maintaining our attention to any opportunities for realization. We still believe in it! Which is where we get the saying "Hope is life".

In fact, hope is not just life, and that is what we are going to see now. For we can, at last, get to the essential message of this book, by prolonging the description of this marvellous mechanism with the help of God!

When we find ourselves in a situation where the conditions and stages we have just described are present, we can be confident because at the same time, the probability of God's help is increasing all by itself.

Because causal opportunities are eliminated!

Because the potentials for abandonment are being eliminated!

Because by keeping up hope, we are helping ourselves to place ourselves in the right conditions, by remaining open!

And the potentials for realization see their probabilities increase.

And then an extraordinary phenomenon occurs!

The nearer we get to the eventual meeting point with a potential for realization, the more that same potential's probability continues increasing automatically. This increase in probability results in the instant "displacement" of the potential towards the encounter with the trajectory of our own life!

From a mathematical point of view, we are dealing with a deviation of ranges with their maximum probability values given that all the potential trajectories continue to coexist.

The displacement takes place simultaneously in space and in time. For now, though, it happens in the future. In the future, it has already resulted in the implicated trajectories coming closer together. And this results in a further increase in the probabilities of these same trajectories crossing paths, and so on and so forth! Which all leads, in the end, to a phenomenon of genuine attraction between the trajectories implicated in the non-causal meeting that in this way creates the miracle, the magic or God's helping hand. And it works, whether the help is expected or not, because the only thing that counts is for there to have been an intention planted in the past, favourable to this type of future event.

It is possible to draw up a proper mathematical formula for this law of attraction, based on a simple rule applicable in our four-dimensional space. When the purely causal probability of an encounter increases inversely to distance in the realized world, this doesn't take into account what we have just described, that the trajectories of life implicated in the non-realized world will also be deviated towards each other in a non-causal way, when the implicated potentials converge. If we take account of this conjugation between causal and non-causal factors, we notice that the probability of meeting increases in an inversely proportional way to the square of the distance. And so we find ourselves in the same situation as two massive objects subjected to the law of universal gravitation*.

Therefore the law of attraction for life trajectories linked by intention appears to originate from the same mechanism and the same equations as the law of universal gravitation.

So when all the conditions are reunited, we would appear to be able to attract to us, as well as being attracted to, the opportunities likely to realize our intentions.

Therefore the law of universal gravitation would not appear to apply only to material objects. It would also appear to apply to life trajectories rendered "massive" by the fluids that come from our intentions. Which is the same as broadening the law of gravitation to apply to our inner space.

The extra, invisible dimensions of our inner space, therefore, would appear to contain a sort of fluid "matter" that submits to exactly the same laws as the matter in our physical space!

I arrive at this conclusion, which in many ways could be qualified as a "spiritualist view", while yet employing logic that is the most rational there can be, in a supposedly double causal universe. Even if this point of view depends on a few short cuts made thanks to a dose of personal intuition and relying heavily on metaphors, it seems to me that the following conclusion will prove "roadworthy": **the Law of Converging Parts becomes a genuine law of attraction for life trajectories when considered dynamically!**

Those who have concentrated on the power of intention (30) and put it to the test are numerous indeed. For decades, numerous researchers all over the world have been interested in this phenomenon, carrying out experiments which, contrary to my own, are truly scientific (25). Most of the time, they are trying to explain parapsychological phenomena such as telepathy and psychokinesis. Generally, however, their work tries somewhat awkwardly to agree with purely causal logic, and this limits their range.

While we are on the subject, it is interesting to point out that the causal approach does not let us explain the fact that we observe absolutely no difference between micro-PK and precognition. Whereas with double causality, not only does a theoretical explanation finally emerge, but it also thoroughly justifies this lack of difference.

However, I must draw attention to the existence of a remarkable alternative theory, the one of the Akashic field of Ervin Laszlo (16), based on the existence of a field of information. It would be interesting to study the similarities one can draw between Laszlo's field and the "supra-dimensional" space of double causality that is home to the Spirit, which is also a field of information, but that is a study that goes beyond the framework of this particular book.

I will not go any further, either, into the reasons why double causality might provide the beginnings of an explanation for parapsychological phenomena, although it seems obvious to me that this potential exists. Similarly, double causality shows even more obvious potential for explaining many other strange phenomena, such as the placebo effect, clairvoyance, drawing cards, prayer, etc. Rather than try to justify this potential, something which would require more books, I will restrict myself to a brief quotation from Arthur Koestler (15) in *The Roots of Chance*:

"One could substitute for the terms "serial" and "synchronicity", with their awkward insistence on on time, the neutral expression of confluent events. Such events appear to be incidences of a tendency to integrate. The appearance of Jung's scarab would seem to be a confluent event, as do psychokinetic effects on rolling dice, and so many other a-causal paranormal phenomena. What appears to make them significant is that they give the impression of being causally linked, something they obviously are not: a sort of pseudo-causality. It looks like the scarab is "drawn" to Jung's window by the telling of a dream that the dice obey the will of the person doing the experiment, that the clairvoyant sees cards hidden from view. The potentials for integration in life seem to include the ability to produce pseudo-causal effects, to produce a confluent event without worrying, if we can put it that way, about employing physical agents."

The pseudo-causality Arthur Koestler talks about is obviously linked to second causality, and its potential for integration, able to produce confluent events, is none other than the Law of Converging Parts, the law of universal attraction that explains all "confluent" events such as synchronicity and a number of parapsychological phenomena.

Here we are then, forced at the same time to recognize all the potential but also the implications at the origin of drifts in the theory of double causality: it would appear to support things we usually classify with the paranormal and superstitions!

Today, however, with the theory of double causality, we can no longer keep turning a blind eye. All the scientific results that have long been accumulating prior to this theory, and which tend towards the same conclusions, that of the integration of the soul into science, will now be able to converge.

Chapter 17—Summary

The different potentials for the realization of our future, represented by the different branches of our Trees of Life, are dynamic: they are instantly modified as time passes, according to our actions and changing intentions.

The result is that when we increase our chances of living a particular event, observation or coincidence, etc. through our actions in the present, the potential corresponding to this event spontaneously shifts in our future, increasing our probabilities for observing it. Our chances of encountering it are proportionally multiplied, because two causalities, not just one, are now working together towards its realization.

This translates into an attraction of life trajectories rendered probable and consequently "massive" by our intentions, inversely proportional to the square of the "distance between trajectories".

The Law of Universal Attraction of Life Trajectories is the dynamic expression of the Law of Converging Parts, which tends to unite like with like. It is analogous to the Law of universal gravitation.

XVIII.

THE EMERGENCE OF THE MIND

In which we reconcile science and the soul

"Henri, I've got a problem, I think the synchro-stimulator has broken down! And we haven't got any recordings, the only ones we made were memorized in it!"

"What? Make sure you get it repaired quickly, you know the big boss from psychophysics programmes is coming round. If your tweaking doesn't work, then the extension of our grant will be cancelled!"

"Ouch! But when I said it had broken down, I really meant it was dead! The circuits have blown! It'll be impossible to repair it for at least a month!"

"Just sort it out if this programme goes down then all the others will be affected. We're already in the red! You're our last hope for carrying with this sort of research. I know you, you'll find a solution for me!"

"Yes, but not this time, really not, I just can't see how, even if a miracle happened... Although... But no, no, really... I'm afraid that..."

"What do you mean? No buts about it, I can see that you've got an idea, just tell me!"

"No, I can't, it's impossible, no one would accept something like that, unless... Oh, but no, it's just too ridiculous!"

"Oh for goodness' sake, just say it, I'm listening!"

"We can't doctor the video, we're agreed on that, it would be deception, so we have to film a real anti-mixing process. So, what I thought we could do was reproduce the process, but without the machine! We'd just have to play the film afterwards."

"What! But you said there were no recordings! Give it to me straight, did you manage to record something somewhere else? Phew, rescue's at hand!"

"No, no, it's not that! I meant... well, let's just call it the old method."

"What old method? There is no old method! I don't know what you're talking about!"

"I know you won't be keen on this, but I'm suggesting we simply inverse the mixing process and call on a real mind!"

"What! We can't do something like that, we'll be lambasted, forget it, you know that if we've managed to get funding up till now, it's precisely because we wrong-footed all the parapsychological labs! Because we didn't allow ourselves to put humans into the machine! And that's your own expression!"

"But we don't have to admit to it..."

"Phew!... Eek, I see what you mean... You want to take us back to the stone age, on the sly... Listen, of course, they'll notice, even supposing you manage to get a ultra-gifted subject, the anti-mixing of his mind will still be much more disordered than what we could have achieved with the machine... They'll smell it a mile off! It's too risky!"

"Don't be so sure, that's a false concept, everyone believes it but it's actually because in olden times, they used a natural source of energy! It's because the subject was forced to take energy from the environment that he generated a disordered anti-mix!"

"Really? But how do you know that? Hey, have you been hiding things from me? How do you stop the subject from having to take energy from the environment?"

"Erm... well... I've got a bit of a trick, but... I can't tell you what it is! Or at least, it can't be published yet."

"What! Go on, you can tell me!"

"No, it's just that I realized something... I mean... I don't know how to tell you this."

"What did you realize?"

"How to do without the machine, without needing to use a gifted subject. You, for example, you connect to the vibrating radio source by touching it with one hand, and you place your other hand in between it and the top of the mix. Meanwhile, I take care of the recording and the rest."

"What! But that can never work! It's ridiculous!"

"Yes, it can, though, on condition that you think about... I don't know... about your grandmother, for instance, and then you just let me get on with it. You know, it's just a consequence of double causality! But no one knows that because they never actually experimented with a series of minds."

Before using double causality to shape a definition of what we could call the Mind, it is important to summarize our theory in a few lines, starting with the new elements that now emerge from the spatial study of the effects of our intentions.

The human is submerged in a three-dimensional physical space that possesses extra, invisible dimensions nonetheless "inhabited" outside of time, in the past and in the future, by a part of their being situated outside the brain. As it is probabilistic in nature, it is fluid in form, and we could even say that it is blurred!

This part of the human is actually the expression of his free will, in other words, his capacity to modify his own life trajectory and incidentally that of others'. The space in which the human can live outside of time translates into all the potentials whose reference point is his current trajectory.

Outside the space that hosts this referential trajectory, the other parallel spaces exist only in the state of potentialities. But the part of us that inhabits these spaces may draw our referential trajectory or "destiny" towards itself in order to solicit a change of space, by increasing the probabilities of a "non-causal bridge" coming to offer us a bifurcation.

This offer can appear in different, more or less conscious, forms, ranging from instinct and intuition for the most unconscious among them, to the most conscious sorts of "signs" such as coincidences and synchronicities.

So it is indeed as a last resort that our free will, under the influence of a deeply held, more or less conscious conviction, decides to follow a non-causal bifurcation, and it is indeed because it is non-causal that there is a choice between two paths, causal and non-causal. This choice defines the very essence of the mind that thus affirms itself. The mind is by definition the one that makes the choice. At each bifurcation, there are two possible minds.

The production mechanism of non-causal opportunities, intuitions, favourable chance, coincidences, synchronicities and even "miraculous" sporting exploits, often popularly labelled as the "Hand of God", originates from an inverted determinism forming a bridge between the future source of our intentions and our near future, the one we are heading towards. It is the mechanism of second causality, and it is only possible because inverted determinism sufficiently "sustains" probabilities as it comes back in time.

This determinism is expressed "statically" by the Law of Converging Parts and "dynamically" by a law of universal attraction of life trajectories linked by intention. This intention is materialized in multi-dimensional space by probabilities which, taken as a whole, are equivalent to a mass of fluid trickling towards the present by travelling back in time.

These fluid objects, whose mass corresponds to the sum of the distribution of their probabilities, are likely to be materialized in a present towards which they are travelling, but only if their mass is large enough to draw our attention. They correspond to pure, omnipresent information. The role of intention is to modify their distribution by provoking a change or displacement of their form.

This hypothesis of the omnipresence of a sort of material information in an infinite whole of parallel universes should not shock our intellect, for all these spaces have no reason for being empty.

This omnipresence performs an elegant role with regards to the concept of planting intention: when we plant an intention, it acts like an instant increase in probabilities, without telling us exactly where the outlines of this planted intention are situated. Now, the omnipresence of the intention can perfectly well "embody" this act of planting, using changes or displacements in form, something which is infinitely more elegant.

Such a way of conceiving things leads us inexorably to the notion of the Mind, while at the same time allowing us to perceive all its properties. Let us not venture too far, however. At this stage, the supposed omnipresence of the result of our intentions, that is, our "being", in fact, the omnipresence that brings us to "embody" our mind in our parallel spaces and outside of time, is still just a hypothesis.

However, what is interesting about it is that it makes us understand the existence of something more fundamental than our intentions: that, quite simply, of our being or Mind. Our Mind would seem to be omnipresent, and when we modify our intentions we modify our Mind, or else we displace it!

So it isn't so much our intentions as our very being, that is, our Mind, that is inextricably linked to our free will.

Thus the mind appears as a consequence of our assumption of free will! And the way we have been brought to embody it in parallel spaces outside of time has been done entirely naturally, using an entirely rational process of reasoning.

The existence of the Mind outside of time would appear to result, paradoxically, in a point of view at once rational, Cartesian, material and determinist:

– rational, because causality does not allow us to explain coincidences, and so we have solved this problem in a rational manner using double causality, which merely reinforces causality,

– Cartesian, because we have called upon a multidimensional, Cartesian space, itself susceptible of hosting an infinity of parallel Cartesian three-dimensional spaces,

– materialist, because our intentions are well and truly materialized in our inner space by a "fluid",

– determinist, because the effect of the intention, symbolized by this fluid flowing towards to past, happens in a determinist way.

This approach to the Mind is therefore a very "physical" one, and while other such approaches exist, they all come within the context of causality (16) (22). So we can hope gradually to witness scientific recognition for the soundness of spirituality, a reconciliation between soul and science, maybe even the introduction of faith into experiments, as our fiction with the synchro-stimulator suggests.

As for eliminating chance from the laws of physics, we are not there yet, and there is a long way to go. Even on the side of religion, it will not be easy to accept the new concepts of the mind emerging from physics, for they can have revolutionary implications.

It is the youngest, more minority religions, like the Bahá'í faith, for example, with a mere 160 years of existence, that are best suited to accept double causality, via the recognition of the fundamental link between science and faith. Here is an extract from Bahá'í writings tending in this direction:

"Religion and science are the two wings that allow man's intelligence to soar high above and allow the human soul to progress. It is impossible to fly with only one wing. If someone were to try to fly using only the wing of religion, he would soon fall into the quagmire of superstition, while, on the other hand, with only the wing of science, he would make no progress, but sink into the desperate foundry of materialism*."

This quotation reveals a very beautiful representation of double causality. This too has two wings, which unfold in two different directions, if we consider that the left wing is that of science, and the right that of religion:
– the left wing, then, is the wing of first causality, in the direction of time;
– the right wing is the wing of second causality, in the inverse direction of time.
Let us pursue this analogy:
– the left brain is the brain of analysis and logic;
– the right brain is the brain of intuition.
So the two sides of our brain are already adapted for double causality. Could we hope for confirmation more seductive than this?

I would like to end this chapter by quoting from an unknown source of metaphysical writing, that present the concept of "tunnel man", something we could consider as a new metaphor for a possible branch on our Tree of Life. Instead of taking the example of a branch, these writings take that of a roll: a cylinder or flexible tunnel. They consider that time does not exist and that a man's whole life can be represented by a long tunnel that can be moved if necessary, by rolling it, for example, in order to modify his destiny.

Imagine a little man buried in time inside this tunnel, moving through it as he progresses in life. While the man moves inside the tunnel, the man's soul, for whom time does not exist, does not act on the body of the man to make him progress, but on the length of the whole tunnel at once, displacing it.

So this theory of "tunnel man" contains two sources of causality. The first is physical, where the man moves through the tunnel by displacing his body, while the second calls on a disembodied entity that has the ability to modify the position of everything in the tunnel at once.

If we consider all the possible positions of the tunnel as a whole, then we find the Tree of Life.

The illustration employed to make us understand that time does not exist for the soul is that the man walks through the tunnel with a lamp to light his way. The furniture ahead of him, in the dark, corresponds to his future, and the things behind him, no longer in the light, correspond to past events. This man's present, then, is his lamp, revealing a step-by-step reality that is very localized, although past and future realities are still there and existed before.

We are then told that, even though the future has already been written, free will is no deception, for the soul put its tunnel down on the earth only once. Before doing so, the soul studied the shape of the ground, the bumps and hollows, hills and trees and rocks that needed to be avoided.

The soul saw the tunnels of other humans being placed here and there. It saw all this from way up in its timeless place, and freely decided upon its path in life.

Another metaphor used in these writings is one where we are looking across at a house through a little hole made in a piece of card, preventing us from seeing the entire house all at once.

Thus the conscience of our body, the one that comes from our biological brain, only discovers the whole house by gently sweeping across the entire surface of the house through the hole in the card, because its field of vision is very much reduced. But the conscience of the soul, on the contrary, sees the house in its entirety, for there is no card held before its eyes.

Mind and Soul seem here to be considered as two words used to designate the same thing. If we believe that, aren't we forgetting that in French the word "âme" (soul) is created from the word "amour" (love)?

Chapter 18—Summary

The geographical localization of the modifications produced by our intentions in our "future space" allows us to identify an omnipresent substance that we define as a Mind, because it has all the appropriate properties described in literature, notably the property for instantaneous displacement: when we modify our intentions, we immediately make our Mind change "place", while at the same time modifying its shape, which encompasses the extent of all the potentials stimulated by our new "state of Mind".

The ability of our Mind, or "being", to move would thus appear to be inextricably linked to our non-conditioned intentions, that is, our free will.

The mind, therefore, appears to be the geometric expression of our free will!

XIX.

THE CYCLE OF LOVE

In which we divulge the magic's secret!

Until now, we have spoken about intention as a function of our free will that allows us to direct it and so determine the future towards which we wish to travel, and we have embodied this destination by representing it in a space fitted with extra dimensions.

Using this spatial representation, we have caused the emergence of the Mind, an omnipresent form altered by intention. Whereas intention remains attached to the present, the Mind is timeless, and this is where the difference lies. Apart from its probabilist nature, we know very little about the essence of this form that can move about in the future and turn itself into rain flowing gently back down the waterways towards the present.

If, for example, we visualize lying on warm sand on a beach lined with palm trees, we can also form the intention actually to experience the corresponding sensations. First the film of these events presents itself to our imagination, then our intention takes hold of it: our Mind then repositions itself in the future like a cloud formation, and before we even know to make it into a reality, our imaginary vision is replaced by a potential already working in our favour by watering new spaces. We have instantly modified the probabilities on our Tree of Life!

What is the driving substance behind this momentum which, by displacing our Mind, has transformed a visualization into a new life potential, even though we have not yet made any plans?

When we revealed the spatial nature of the Mind, we set aside the question of the nature of its substance, this "essence" circulating among the corridors of time.

Using an analogy with the fuel ("essence" in French) that circulates in a combustion engine, making it run by turning fuel into mechanical energy, might there also be some sort of "fuel" or essence powering the engine of our free will?

There is indeed an "energy" transformation at work in a non-causal cycle governed by intention. The fuel of desire is inflamed by intention, instantly spreading its "emotional energy" into the future, then this energy travels back in time towards the present and emerges somewhere in the form of a potential for realization: the fuel actually appears to have been transformed during the cycle!

This cycle has one particularity, however, in that it needs a certain amount of time to lapse. The circulation of emotional fuel is characterized here by a critical and uncertain phase: waiting for realization. During this phase, everything may be thrown into doubt, and the circulation may stop: the engine may stall!

We will now list the different phases of the cycle. In our model of the Mind, we have already observed the first phases, starting with the phase of **desire**. Before we form an intention, we are generally in the phase of desire. The difference between desire and intention is that intention is not satisfied with this phase, because it wants to make the object of desire enter into reality. Desire creates and sustains the fuel, but the fuel hasn't yet been set alight or even displaced outside the tank. For desire keeps fuel anchored in the present, like a parked vehicle keeps its fuel in the tank. In order for fuel to circulate, the vehicle must start its engine and so lose some of the fuel in the tank.

Desire cannot really imagine that the intention might be achieved, otherwise it would be more a question of anticipatory joy. On the contrary, it is often accompanied by frustration. And so the more conscience is aware of how difficult the thing is to achieve, the more desire grows in proportion: the starter isn't working!

However, once we stop holding the object of desire prisoner inside the fuel tank, we achieve an attitude of *detachment* that allows the fuel of desire to be drawn through into the phase of the **wish**. Now the starter can begin to turn.

Wishing, however, is like an intention yet to be born, because it hasn't yet formed its resolution, nor moved in any direction at all. The vehicle remains stationary. Wishing cannot make us change branches or move our tunnel. It's like an intent of intention, a promise of intention. The wish doesn't quite know where it ought to go yet, nor even which vehicle it should use, for this depends on several factors: wishing might find itself without any car at all.

We can change vehicles, or even have several, just like we can have several wishes concerning our next holidays, for example, but we cannot really drive more than one vehicle at a time: we can only have a single intention, for intention, unlike wishing, consists of action.

We had better be aware of this, because having several wishes, just like possessing several vehicles, is useless to us until we understand that in order to move, we have to detach ourselves from our doubts, stop hesitating, and nourish a genuine intention: we have to choose a roadworthy vehicle and start the engine.

Once the vehicle has started, the wish becomes an **intention**, because we leave the car park. Careful, though! From now on, we need to take into account the lie of the land! We must pay attention all the time, ready to make an observation: now we are entering the phase of **attention**.

However, we still need to have really left the phase of intention, by detaching ourselves from the route shown on the map — from the means, in other words. Now that we are on the move, the mind must leave the map behind, it has to let go in order to allow us to concentrate on the ground we are covering in the present.

Everything gets easier in the attention phase, because all we have to do is drive to arrive at our destination without trying. However, one difficult remains: we don't know exactly which road will take us to the right destination, nor how long the journey will take. We are in the critical, uncertain phase, and this is why we need **faith**.

The faith phase would be useless if intention could find a causal means of achievement, that is, if we knew the road by heart, in which case the journey would not be a result of second causality. Faith allows us to keep driving even if we don't know the road. It allows us to become detached from being excessively attentive out of ignorance.

We have to remain attentive, in fact, but keep driving normally. Faith allows us to see no re-emergence of excessive doubt, or any hesitancy followed by doubt, which would cause the engine to stall. However, as we have already seen, a certain dose of doubt reinforces faith. Excessive doubt, though, like telling ourselves we are lost, would be the surest way of stopping and making a U-turn or seeking a causal plan.

This is why the best mental state for supporting and engaging second causality is "faith", for it has to be maintained over a period of time without undoing the intention. Not religious faith, but faith in oneself, or if we prefer: faith in God's help.

But what is this word "faith" that sneaks up to take the place of our "intention" and maintains itself over time?

We need faith when we don't know how we will be able to realize our intentions. We have already explored at length the reasons why faith might be favourable to second causality. Our conclusion was that it allowed us to increase our chances, by acting as a barrier to first causality and the temptation to give up.

So faith appears to be intention's firmest ally when no causal solution can be seen for its realization. Careful, though: faith becomes useless when it causes our doubts to disappear, and once again, it is important to know how to be *detached* from it.

For faith, in turn, must hand the baton to **confidence**. Having managed to chase away our doubts, we may well be ready to encounter signs along our road, but we still need to know how to decode them: how to trust them. For if faith takes us towards such signs, we need to trust the decision they suggest to us: and generally, it is the first impression that counts.

Once our confidence has allowed us to make the right choice at each bifurcation, we finally enter the phase of **"aspiration"**: once we have confirmation that we are on the right road, we can feel that we are going to reach our destination. We have noticed a change in the scenery, and that it is starting to look like our destination. We are drawn towards it.

But what characterizes the destinations of our intentions? Happiness, pleasure, joy? Not necessarily, because we might have the intention of accomplishing some courageous or risky act, placing us in a difficult situation in order to help someone, for a noble cause...

If the object of our intention isn't to create something beautiful, good or positive, or to find happiness, then isn't it to generate all that for someone else?

Why turn desires to wishes, wishes to intentions, intentions to attention and then to faith, confidence and aspiration, to head for a destination that would create nothing, bring no joy, either to us or to anyone else?

What would the meaning of the execution of a cycle that brought no good to anyone? Wouldn't it be better, in that case, to keep the emotional fuel inside the tank, and wait for something better to come along?

Consider the transformations that take place in the emotional cycle: the fuel of desire turns into wishing, then intention, then attention, then faith, then confidence, then aspiration, until it finally comes back to us in the form of joy, when we reach our destination!

Note that the completion of desire into joy undergoes two processes, that of the gift of oneself in order to keep the cycle going, and that of the detachment: we have to give up each state (of mind or soul) in order to welcome the next, thus allowing the fuel to be transformed in order to modify our potential futures.

Then the cycle is completed and the fuel returns in the form of joy. Its disappearances were fictitious ones, the magic of life was set in motion via the future. The fuel in a combustion engine disappears in the same way by being transformed into mechanical energy. In the engine of our free will, the emotional fuel disappears to turn into potential, completing its mechanical work in the corridors of inverted time.

Let us examine the phases of the cycle in which emotional fuel has not yet disappeared by leaving the present, and has then started reappearing: desire, wish, … then aspiration and joy!

To this, we could add all the different expressions of joy: happiness, pleasure, satisfaction...

What is the common factor essential to each of these requirements?

Couldn't it simply be everything that justifies our love for life? In which case, couldn't love itself be their fuel?

Couldn't love be the fuel for the engine of our free will, the fuel that frees us from a conditioned existence, the fuel that strives for our genuine liberation?

Couldn't love be the fuel that powers our freedom? We know that this is about our freedom, for such is our fundamental assumption. Without freedom, there is no theory of double causality, no engine of free will, no need for fuel!

And anyway, after all, what can we hold up to an objective as noble as freedom, if it isn't love? What is there as noble as love, if not freedom?

And it is quite obvious that the request for love corresponds perfectly to the demands of desire at the beginning of the cycle, and that once the cycle of love ends, it is fulfilled by joy. It may be only partially fulfilled, by new hopes and desires, but the cycle of love is always ready to start again...

Even though it is composed of fuel that can alter its appearance, just like energy, in the end love is clearly linked to the fluid substance that infiltrates our potential life trajectories in order to help us realize them.

We have already seen how the mind defines their contours, and that we are dealing with a spatial distribution created by our intentions.

If the mind defines the contours, what still needs to be defined?

If you ask any physicist, he will reply: what still needs to be defined is the amplitude, the volume or even the energy. But what do amplitude, volume and energy mean for an object situated in space?

It is quite simple: an object's volume, amplitude and energy are directly correlated to its mass. In this case, though, shouldn't the mass be subject to universal gravitation, like any self-respecting mass? Well, yes, that is exactly what happens, and it is a consequence of double causality: the law of universal attraction!

Thus, if we had to give a physical description of love, it would be the energy of the mind when producing its effects, but also its mass when kept in a resting state.

It is interesting to see that when love turns into energy, we no longer recognize it as such. Intention, attention, faith, confidence: so many expectations that it will come back to us from the future.

However, when it is expressed in the present by its mass, could love be composed of particles? What should we say about gravitons*? Because we can't yet detect them, men are busy building accelerators of quite gigantic proportions in order to try: is this so surprising?

Whatever the case, particles of gravity love, if such a thing exists, evolve in different dimensions. Coming from the same law, does this also mean we can deduce that they are identical?

We can always try: then love would be nothing but the energy of the fluid substance that spreads through our inner space and constitutes our mind. So what happens when it travels back down into our physical space?

Now let us consider what we know as coming from a reliable source: the source where it becomes liquid again, the source whose amplitude it defines.

What can we say about love then? Considering everything we have just seen, it is easier to start with all its results that are not actually it: it is neither love desire, nor love joy, nor love pleasure, nor love satisfaction, nor even love happiness!

For all these sorts of love are just the results of a love that has flowed onto the earth in reply to our intentions, and that is not how we should define love, real love, the love that springs from the source.

For we are talking here about love as a pure source, not as a substance that has been planted, that has already interacted with space-time, that might become impure.

Just as our intentions need to be purified in order to separate genuine from merely illusory free will, doesn't our love, too, result from the purification of all our feelings?

In order for our mind, whose amplitude is defined by love, to be able to separate itself from the part of us that is situated in the present, shouldn't there in fact be some sort of purification?

Thus genuine love would appear to be discrete and invisible. We would only be able to observe its effects. Purified love, transmitted by intention, freed through detachment and energized by the gift of oneself, would appear to act in the future and show its effects in the form of Grace. Love received might not know what its origins were, any more than we know the origins of chance.

Disappearing into the future, love would be transmitted like light, moving instantly like a beam.

In the opposite direction, however, love would become solid once again like a fluid, like water flowing back towards the past to bring beings together.

Where does this difference come from, these two quite distinct expressions of love, at once fluid and light energy?

Only an effect that is much slower to appear can be transmitted to the present from the swollen, amplified source energized by love, for love has to cross a space-time involving numerous life trajectories.

The slightest modification of intention requires readjustment of all the potentials separating the present from the future, because several Trees of Life are concerned and they are all tangled up with each other.

When our life potentials are modified, those of all the other people involved in our lives are too, and all the potentials of all the other people that they are linked to in their turn, people whose lives appear at first to be quite independent of our own. Some apparently minor changes, if propagated at the speed of light, could generate veritable waves of probability throughout the whole of humanity, and then we would have to wait for all the potentials thus propagated to settle down, before they could stabilize.

It becomes doubtful that these sorts of waves could ever become stable, if we consider the cumulated effect of all the intentions of all beings.

Is this really how things happen?

Fortunately not, for as they travel back in time the potentials have to wait before they can spread out, because they depend on the trajectories of all the beings involved "in their present". A single intention is not enough, it must be confirmed by adequate active effects. Present changes created by intention thus gently reinforce potentials, and this results in them growing at the speed of passing time. We have to help ourselves, don't forget. For intention can only really be potentialized if the lie of the land permits. No one has to achieve the impossible. Everything that is possible will be done, but time has to do its work, which is to acknowledge all the bifurcations.

And this is how, at the start, love is like light or electrical energy, instantly propagated. But make it travel across the lie of the land and move over matter in the opposite direction, and just watch the results: love circulates slowly.

To understand properly the effect of the lie of the land, ask an electronics engineer how he would go about stabilizing an electrical circuit, a source of power, for example. He would explain that you have to give the circuit a capacity or even a resistance, to absorb power surges and compensate for drops in voltage, thus eliminating undulatory phenomena, evening out the flow of energy and slowing down the circulation of electricity.

And then the circulation of energy becomes fluid!

When in the future a wave of probability, directly fed by love accompanied by intention is propagated into the past, the same phenomenon occurs. The land, consisting of a multi-dimensional space-time separating the present from a future to which it must be linked, starts playing the role of a circuit fitted with resistors and capacitors.

The circulation of love, which starts off as instantaneous, like light, ends up circulating in a fluid way, like water.

And this is why, even though love is emitted like light, it is transformed as a fluid when it spreads its magic over the earth.

So this love is nothing other than the fluid that prepares our life trajectories by linking the present and the future. Without love, we head towards the unknown, spend our time playing our joker, and in this case, fear appears on the horizon. But if love comes back to us, then fear vanishes, and we can once again play the game of our life, and seize our chances.

Thus, when it "comes back" into the present, after having percolated in the space-time that separates us from our future, love turns into fluid that can fill our tank, but only on condition that we are ready to receive it and seize our chances. If we are afraid, our tank stays empty.

Each of us has his own tank containing a fluid of vital love, necessary to the proper working of his psyche. This love-based fluid makes each person's life easier, according to how full our tank is. The fuller it is, the more that person will benefit from the magic of life, not only attracting chance to themselves but also drawing everybody else along in their wake. For this fluid that attracts life trajectories outside of present time by synchronizing them, also draws them into the present, without any magic being needed, for it is so obvious that in the present, love attracts love.

"Lightning" love illuminates but cannot flow immediately, without having crossed space outside of time. At best, it can be transmitted directly onto the earth, but only if the person receiving it has been illuminated and is present within its energy field. If they move out of the field, the love disappears. Luckily, we can predict that if they remain present, then love will eventually rain down constantly around them. The fuel tanks of everyone present can then be filled so that love subsists outside the energy field, making their fears disappear in a lasting way. But aren't we in some sort of dreamworld here?

In order to be transmitted gently and in a much surer way, love moves by flowing. It is an expansive movement, whether it flows towards the outside or towards the inside, melting the boundaries between inside and outside.

Letting go, acceptance and detachment are all contained within the nature of love, for they are the wings of second causality. Fear, on the contrary, feeds on first causality, on control and calculation, how to receive or draw towards oneself, and this is why love is more present in the act of "giving" than in the act of "receiving". Fear contracts, degrades, imprisons and impoverishes, the opposite of expansive, restoring, open and generous love.

Love enriches itself. It receives what it gives, sometimes a hundredfold.

Love received increases its ability to illuminate and fills up the sources, making it flow with all the more intensity, and this is an infinite, virtuous circle.

Thus Love enriches itself all by itself, and can enrich each new being as it grows and encounters other beings to love. It tends to expand. And so it tends towards universality.

Love is not a true quantity, because it is always growing. So it cannot be exclusive. If one love excludes other loves, then something is stopping it from circulating. Love also displays a quality: not only does it not diminish when it attaches itself to several people, but it cannot be divided or split. The Love felt for one person automatically spills over onto the others.

Love does not dwell in the ego, because the ego dwells in causality and in the individual. Love resides in acceptation and universality. It can only flow if the ego disappears. If someone is full of ego, Love disappears exactly as second causality disappears when first causality imposes itself. Ego bars the way to love, because it swells up with causality, in the form of personal merit and quality.

Ego says: "I am the source or the origin of this or that! It is thanks to me that such and such has appeared! Thank me!"

Love can only flow if we need it. Ego has no need of Love because it stands as a rival to the source of Love. Love comes from second causality, ego from the first.

Love and ego cannot converge.

That is why, even though Love is so powerful and so necessary, we cannot understand why it is so rare. It is as fragile as a flower. Like a flower, it must be protected, strengthened, watered. Only then can it grow.

A certain amount of time needs to pass before the flower can grow, which is why the effects of Love are not immediate. To witness its effects, we have to water it and water it, give and give again unconditionally.

And then one day a rose appears and our home is filled with its scent. This is how Love appears: always unexpectedly, like a gift of nature.

And because love is impersonal and universal, it cannot illuminate or flow freely until we understand that our being is not limited to ourselves, but encompasses the other. So we must love each other.

And who teaches "Love one another?"

We can see with a certain amount of astonishment that science may well end up teaching us this, too! For this to happen, though, water, and more surely love, has to flow beneath its bridges, and the message of this book has to break down the last bastions of resistance.

Chapter 19—Summary

The Soul is composed of a form—the Mind, and a fuel or essence, which is Love. Second causality operates according to a cycle that makes Love circulate, a cycle that has amplifying virtues during the transformations at work in the future, and produces results in the form of rain in the present.

Whichever way we break down the cycle of Love into phases, they all reveal processes of giving oneself and of detachment, processes that may give us the false impression that love is leaving us.

For the gift of self-causes the expulsion of the fuel of love into the engine of our free will, while detachment, at each phase in the cycle, turns it into a new propelling energy that models our future according to our intentions.

XX.

TALKING TO THE MIND

In which we broaden the field of application of double causality.

The Theory of Double Causality (TDC) was born to explain the magic of life, the magic that makes the world such a small place, by placing along our path the encounters we need in order to evolve. This magic goes further, though: if only we can develop a certain receptiveness, the universe becomes a mirror for our soul, a magic mirror that not only reflects our states of mind, but that informs us about our chosen futures.

However, according to the Mind that I questioned, TDC's potential is even vaster. Here I transcribe his answers to my questions on the subject. To incite him to respond, I first asked him whether my book corresponded to his request, by daring to inverse our roles. This is how I came to see him smile, and how I managed to read his answers in an inner smile.

"In your opinion, could double causality explain the placebo effect, or even spontaneous healing?"

"Yes, he answered, and the difference between these two cases is merely a question of the amount of energy supplied. Take healers: you can see two processes at work. On the one hand, there is an intention to give, an intention of love that in some way visualizes the healing of the patient, and, on the other hand, there is a supply of energy, through the laying on of hands, for example.

Now imagine the same process that led to the illness in the form of a degradation of cellular organization being repeated in the opposite sense to time, thanks to the effects of this generous intention. You can understand how this might lead to healing, but that an important condition has to be fulfilled. Since we are dealing with an irreversible process, in order to reverse it it is absolutely indispensable for it to draw energy from its surroundings, the same energy that was released when the degradation occurred. But since it cannot be the same energy, because the healing occurred in a different environment to the illness, some other type of energy supply has to be provided. And what purpose does the laying on of hands serve if not to provide the necessary supply of energy?"

"What about energies that aren't really physical and also circulate through hands or by whole-body movements, like Chi energy in T'ai chi ch'uan? Do you think that this discipline has anything to do with double causality?"

"Remember the sessions of T'ai chi ch'uan you went to, and the film they showed you about the exercise of the discipline, demonstrated and commented by Vlady Stevanovich (29), the master who teaches his own method. During his impressive demonstrations, you really had the impression that the man was controlling the circulation of some actual vital fluid, a circulation that was very slow, but very harmonious. Vlady himself described what he felt as something close to love and the giving of self, and it was obvious that this fluid carried balance, harmony and beauty. As for knowing whether this was the "essence of love" that you present as the fluid of second causality, think about it, it is something you have to discover for yourself. I encourage you to study this discipline, for it concords with your hypothesis that such a fluid might intervene in perfect sporting gestures, like the secret art of archery. You are indeed in the presence of perfect gestures here; you only have to watch the master to see it. And there is something else here in common with your theory: emptying the mind! You have to recognize that we can strongly presume here in favour of second causality being at work!"

"Do you think that double causality could explain parapsychological phenomena such as psychokinesis?"

"Isn't that obvious? For instance, remember the "Peoc'h's chicks" micro-PK experiment. Initially conditioned from birth to consider a suspended mobile, called a tychoscope, to be their mother, they managed to influence the mobile's movements after having been shut up in a cage, still within vision of the mobile, and even though the mobile was directed by completely random movements. In fact, it approached the chicks twice as often as it should have done, something which is extremely improbable. The scientists conducting the experiment invoked the influence of the chicks' conscience over the mobile, but it is obvious that what is at work here is an intentional process due to second causality, that operates thanks to the emotional bond, a vector for love between the chicks and their mother.

You have also noticed that a great many experiments have been carried out all over the world using random number generators (RNG), showing the apparent power of intention over matter. Wasn't it so successfully summed up in that book you read by Lynne McTaggart (30), that gives such an excellent review of scientific work related to the science of intention? She puts together an admirable synthesis, even showing the steps to follow in order to learn how to transmit one's own intentions, or make requests of the universe in order to attain one's objectives. She herself concludes with the necessity of being compassionate and empathic towards the people concerned by your wishes, of being optimistic and confident about the subject of your request, of looking for the most agreeable conditions possible for transmitting it, and of knowing how to detach yourself from it.

You have understood that it is not a question of a power of thought over matter, but simply the projection of an intention onto the future which, along with the vector of love, reinforces the potentials for favourable chance. Be glad that there is nothing there that could result in something that would violate the laws of physics."

"Can you confirm that all methods of divination, clairvoyance, tarot, I Ching, can be explained by second causality?"

"Yes, of course, insofar as you only see in those media whatever you project onto them by your intentions. But the upheaval of information that you provoke in your future brings additional information back to you in the present, which is where the sensation of magic comes from, but you must take care not to attach too much importance to them."

"Even astrology?"

"Yes, even astrology. In more ways than one, astrology exploits second causality. Not just through the technique of interpretation, which amplifies the mirror effects using the same mechanisms as the other techniques. There is a genuine synchronization between the trajectories of your potential lives, such as they are already programmed in your futures, and the planetary configurations of your solar system. The Law of Converging Parts creates order, and so it creates a harmony between your environment and the behaviour on your part that it needs to predetermine your life trajectories "in your absence", whenever these happen to be indeterminist. For it knows nothing of chance. As long as your intentions don't influence it—as long as your free will remains illusory—you experience states of mind or encounters synchronous with these configurations, something we can label archetypes. Rest assured, though: all this disappears as soon as man exerts his free will. This is why it is a pity that current astrological practice would have us believe that stars exert a genuine influence, when this influence has no physical reality. Nevertheless, it has to be pointed out that some astrologers understand this, like your friend Michel d'Aoste (38) for example, who puts forward a highly persuasive theory that rules out any physical influence."

"So if I understand this properly, as long as we remain conditioned we are affected by an influence from the stars which isn't an influence at all, but which is simply a synchronization between our behaviour and the dance of the planets?"

"Yes, insofar as that behaviour remains subject to sufficient indeterminism. Because the Law of Converging Parts cannot work except to fill in for a lack of causality. If everything about your behaviour is planned in advance and is subject to no vagaries, then it will be determined by the past."

"But then doesn't that imply that a humanoid robot, conditioned by its very essence by its programme, could be affected by the influence of the stars, on condition that we introduce random behaviour into his programme?"

"Yes, as long as the random behaviour is truly indeterminist. But in practice, man hasn't yet managed to design such a robot. No programme allows us to introduce indeterminist behaviour, and it holds no interest for us, unless it is done to induce free and controllable behaviour. To do that, we would have to introduce the indeterminism into the robot's brain, but in a careful enough way for it not to generate any incoherent behaviour. Your technology is still a long way from achieving such prowess. First, man will have to discover within himself the key to his own indeterminism, the one that allows his brain to be controlled by his Mind."

"And yet, some physicists, like Penrose (23) have produced a theory about that, explaining how quantum effects at work in the brain could bring the necessary indeterminism to the conscience for the emergence of the Mind!"

"Between this way of understanding things and applying it in practice to give your humanoids a conscience, you have some serious progress to make! Not to mention that you have no idea about any procedure that would let you ensure the credibility of the intention displayed by a robot, even less of realizing the function of his free will by means of observation."

"Talking of which, TDC contains an original position on the famous question of the role of the observer in quantum physics. According to the theory, the observer's conscience has no influence over the state resulting from the collapsed wave function, and yet, it is thanks to this conscience that we make our choices, and that new realities are created! Can you enlighten me on this?"

"Your humble servant! To understand it fully, consider yourself in the middle of observing something, like a beautiful landscape or a pretty woman. There are other things you can observe in your field of vision, but you have chosen to focus your attention on these particular things. You realize that you are not transforming anything, your conscience is not acting on the things you are observing, all you are doing is making them enter your own reality, together with everything they inspire in you.

Now suppose that you observation is timeless, meaning that you can see everything, including your own potential futures. So now you will focus on those that particularly interest you, and you will have to be extremely attentive, because you don't observe just with your eyes, but with your free will. The attentiveness you must have comes from your power for bringing the choice suggested to you by your observation into reality. Don't forget that behind each observation, a path may have been created to provide you with the opportunity of realizing your intention.

But you remain free, for you have three possibilities. First, you may not even notice the opportunity you are offered, you may not be paying attention, even though it is showing itself to you. In this case, everything will happen as if it had never even existed, and what will be realized afterwards will be the potential to which you were already heading. The other will vanish. Your observation will have had no effect on reality: you weren't alert enough.

The second possibility is that you observe this opportunity, and that you grasp its meaning, but you take no notice of it. In this case, your observation will have made it enter reality, but it will be ignored and considered as a chance observation, a coincidence, for instance. You may understand the significance of it, and even identify the keys, but refrain, or simply not dare to approach such and such a person, or trust your interpretation of such and such a sign, etc. Very few things will be changed, for you won't have disturbed the universe very much, you will only have triggered a few temporal shifts, brought certain trajectories a little closer or further away. The rest of your life will be identical to the one you would have had if you hadn't observed anything at all. The few synchronizations of events that were provoked will fall into oblivion. At best, you will give them a passing nod, putting off until later the changes they have suggested.

The third possibility is the one for change. It is accompanied by an act of awareness, followed by taking into account an event, a sign, an opportune encounter or anything else likely to change your life. Then you will have instantly modified the structure of the universe, and for a great deal longer than these few moments of observation.

All this, without doing anything but observe, then act, quite naturally!

Now do you understand the miracle by which your mind can create reality?"

"Quite edifying, I give you that. But I wrote that conscience had nothing to do with changes to reality. Aren't I contradicting myself? Won't people say that I've given too much weight to the hypothesis that the observer's conscience plays a role in the result of observation?"

"Anyone who says that will have missed the subtle nature of observation. Conscience has nothing to do with it, because it is an effect, not a cause, of observation. It is merely a by-product of your brain activity. Your awareness only comes after your alert mind has made an evaluation of your observation. It is your being, and its "openness", that will have allowed this evaluation to follow its course if it progresses to completion. So you can see that your conscience has nothing to do with it, and even that your mind can remain passive, like a computer, in this matter. The only thing that counts is this: are you, or are you not, going to prepare or make a choice suggested to you by evaluating an observation, in order to orientate the rest of your life? If the answer is yes, then the awareness of your choice merely presides over the creation of a new reality. Your Mind will simply have selected it from among all the other possible realities. It will not in any way have created it, since every possible reality has already been created in advance by the universe.

And take careful note: most of the time this process takes place unconsciously, and it is called intuition."

"So if I'm following you correctly, the case in which the answer is "No" is a very general one, in a society where we are completely ignorant of the fact that chance can be something other than a result of luck, am I right?"

"That's not entirely accurate. Many people believe in their luck. Even if they don't know how it works, they know how to make use of their opportunities and don't ask themselves too many questions about how they come to enter their lives. Which is lucky, really, because if they thought about it, they would risk finding themselves in a causal operating mode, which would cause the entire thing to collapse."

"Ah, right, there we are, I wanted us to discuss this. What do you call a causal operating mode?"

"The causal operating mode is, after a certain fashion, a brainwashing process of our modern society. It is only with such a mode that we can adapt, because we would risk rejection were we to start imposing the complementary mode. The causal mode consists of acting in a reasonable sort of way in your own life, nothing more. It consists of systematically planning, analysing, calculating and respecting specific stages, in order to achieve a particular objective. Knowing in advance everything you will need, preparing your itinerary, making a plan, learning to express yourself, reading the user's manual, etc. Everything, absolutely everything you do in life when you work or make plans is set down within the framework of causality, even if only to obtain other people's help. In fact, it is impossible to escape from this operating mode, except in a very piecemeal sort of way."

"So how do we escape from it? What would be a non-causal way of operating? I suppose that if one doesn't obey one mode, one obeys the other?"

"No, not exactly, you can obey both at the same time, and in fact this is recommended in order to find a good balance. The non-causal operating mode consists in, first of all, raising your level of awareness, alertness or vibrancy, if you prefer. It's a case of listening to your surroundings, but be careful: you mustn't be actually trying to listen to anything, otherwise you'll fall back onto a purely causal focus, into reflection, thinking about problems, etc. We can't trick ourselves. We have to be "natural" if we want to be able to perceive non-causal opportunities. Now, it's easy to understand that it's better not to be stressed, or haunted by things you have to do, or overtaken by stereotypes or "overprocessed ideas". Ideally, one needs to be relaxed, disinterested and thoughtful, and exploit every margin of freedom we have in order to stay attentive to our surroundings.

But these are only general conditions. There are cases where you are automatically placed in a non-causal operating mode, such as when you experience raw emotion, for example. Well, apart from that the rest is partly a question of practice, because knowing how to seize opportunities means not falling into the opposite trap, which would be to see them everywhere. It is like practising a sport that requires skill, like boardsports, for example, which are a good illustration."

"Really? Why are they such a good illustration?"

"Because you have to know how to "surf" along your life. Surfing means staying above things as much as possible, or being alert, so that you don't fall into the trough of the wave, otherwise you have no choice. And the advantage of this sport is that you don't have time to think about whether you should do this or that. You have to act instinctively every single moment. Instinctive or even intuitive acts, in fact, depend on a non-causal way of functioning. With each opportunity, it's the first instant that counts, and what comes immediately to mind is exactly the thing you should follow. If you think about it too much, you miss the slot. That being said, you also have to know how to wait, but without being impatient. For good opportunities are by their very nature unexpected, and you can't choose the moment they'll come along. So it is better to forget about all that and live a normal life, but an alert one, while at the same time escaping from the everyday as often as possible, and keeping in mind that, life being magic, there is no reason to worry about whether you will manage to get out of this situation or that, or how you will do so. The non-causal mode automatically brings you solutions if only you ask for them."

"I brought in the mind, then love, then the soul as if this were a logical succession of consequences of double causality. Thus constituted, is the emergence of the soul obligatory?"

"It is a model, and it is not up to me to evaluate it in your stead. The question of knowing whether you are on the true path is not one that even needs to be asked: your model has its usefulness and must be left to do its job. You have identified the "soul" as the part of the conscience that can escape from your brain, and you have given it substance by according it geometry and amplitude. Your model is simple and coherent, for Mind and Soul, those two fundamentals, allow you respectively to embody intention and its probabilities for actualization. This is why they have to be a product of double causality.

With a model such as this, you are, however, in agreement with spiritual writings, which will bring you partisans and detractors too. For you are moving away from minimalist reserve. It makes me think of a property of the afterlife (10) (19) which authorizes instant displacement of the mind to any place, under the direct influence of will, or of thought. It also makes me think of this feeling of omnipresent love attested to by those who have had near-death experiences, when they are accompanied into the light of the after-life and come out of the tunnel.

I don't encourage you to make justifications for the afterlife, because you are aiming for a minimalist position, even though you are right not to ignore what is relayed to us, when it conforms to what can be immediately deduced from double causality.

The mind's displacement in this space, where your potentials are memorized outside of time, is quite similar to the way your imagination moves outside time when it is making projections into the future. From within the four-dimensional space-time that limits you, however, you can only catch a very blurred glimpse.

Ask yourself whether your dreams are not, after all, your Mind coming back from visits. Isn't it quite logical that these return journeys on its part should be incoherent, because they are generated by its displacements into parallel spaces, outside of any causal succession inflicted by the passing of time? Doesn't this explain the strangeness and incoherence of your dreams?"

"I say that the nature of the vital fluid that intervenes in second causality is that of love. What do you think about that?"

"I can see at least two fundamental reasons for that, apart from the "down to earth" explanations you have given for the cycle of love. The first reason is that we are dealing with a fluid of universal attraction of life trajectories, and the second is that this is a substance common to everyone, which tends not to differentiate between people, rejects the notion of individuality and above all, of ego. Love is what makes you the same as your reader, or makes you identify with him by making every possible effort to anticipate his thoughts. In this way, your thoughts can evolve through people other than yourself, meaning that other expressions will be defined, each time corresponding to the very best that they can give. For you must all understand that you are all part of a single organism, that of God (32), an organism for which you are all simultaneously receptors and actors.

Do not forget the most fundamental thing, though: love is associated with freedom in the functioning of the engine of your free will!"

"In your opinion, how can understanding double causality modify our behaviour in our everyday life?"

"It already affects it through beliefs and religious practices. It is true, however, that working magic daily is quite another story. Man has a great deal of progress to make before being capable of changing his way of living so that he can be sure of maintaining in his everyday life the mental attitude required to provoke the effects of second causality. This means knowing how to "vibrate" at a higher spiritual level, something more or less incompatible with your situation as modern slaves to consumerism.

This magic does work in your life every now and then, though, when emotion and love are both present and you find yourself in a period of great change.

However, I would like to point out a way for working magic every day that is very easy and has to do with second causality, something all of you, numerous as you are, have a tendency to forget, although countless messengers have repeatedly told you."

"Really? I don't follow. I've missed something. Aren't you contradicting yourself? You have just told me that generating the effects of second causality requires knowing how to vibrate at a higher level!"

"I was talking about modifying the future. Now I'm talking about modifying the past. This is quite a different matter, because it doesn't require us to "vibrate" in the same way as for sending an intention into the future."

"Do you mean that we can send intentions into the past? But even if that is the case, I don't see how that could be any easier, on the contrary, in fact, because I find it hard to see how we could modify the past!"

"No, we can't send intentions into the past, no more than we can modify it, because it no longer exists as such! Only its traces are still here, and the past is a determinist result of them, through being recreated from them automatically, instantly. You must understand that by erasing its traces, the past reconstructs itself and changes, putting order back into it! And that is where the true meaning of irreversibility lies!"

"The past reconstructs itself! But that's incredible. How can it do that without us, without observers, if we are co-creators of the universe?"

"But we're not, this co-creation only appears to be as such. Man creates nothing, he only chooses what is created, and even then, only in the direction of the future. In the direction of the past, the universe constructs itself all by itself, without the need for observers. Otherwise, how could you explain the creation of the sun, at a moment in time when no one could have been there to watch?

Listen carefully: a few years ago, you fell and hurt your knee, but you don't remember it. Since nobody saw you fall or complain, and since your knee has completely healed, the past as reconstructed by your Law of Converging Parts has "passed over" the fact that you fell! In this new past, you never even fell at all, and my reminding you of it now isn't a trace to be taken into account, since I am not an actor in your world."

"But how can that be possible? How can a past that has indeed been lived not be memorized by the universe when it memorizes an impressive number of futures that haven't yet been lived?"

"Don't forget that every time you change futures, you change universes! The mere fact of going from one parallel universe to another results in everything changing, future and past simultaneously! Except that the past obviously can't be changed, as long as we make it justify the traces of it we conserve in the present!

As for the future, it is completely different since it is a question of potential lives. The fact that they are multiple doesn't change anything!"

"This is astonishing, because I would have thought that the past and future would be more symmetrical, according to double causality, even though it appears that we don't "pass back" over the pass. But anyway, I don't see what this has to do with the daily working of magic!"

"You don't? What happens in the future with second causality? You water it with love, but it is complicated, you have to respect a certain cycle, and show detachment. Whereas to water the present, nothing could be simpler. All you have to do is do good around you!"

"Really? Well, that's good, I approve! You mean that this good will flow back to the past, everywhere it's still possible to change something, so that it... Goodness me, what could it make happen?"

"It will help to erase negative traces, that's all! It is simply a consequence of your Law of Converging Parts that tends to bring order and simplify everything! As soon as a trace of the past is erased, then the past is instantly simplified! Everybody knows that love helps to forget the past! But what people don't know is that it is able not only to erase it, but above all, replace it by another, much more advantageous past! For by flowing into the past wherever the past can be modified, not only does love bring simplification, but harmony and healing!"

"What? I've never heard the like! But this would mean that pathological liars were right! Because they are specialists at forgetting traces of the past and replacing them by whatever they want, in order to deny the harm they have done, for example! This would justify lying!"

"Not at all, quite the opposite in fact, for several reasons. First of all, lying to oneself is the best way of not forgetting the past, because it remains deeply memorized in the subconscious. Secondly, lying is by definition a negation of traces of the past, otherwise it would never be considered as such by other people. On the contrary, as a result it incites the person who detects it to make sure not to forget anything. So in the end, no one forgets, so lying is the worst possible way of forgetting the past, just like fear, in fact. Fear is founded on the idea that a negative past is going to happen again. Once again, it is a means of preventing traces of the past from being erased."

"So if I understand it correctly, the best way to erase traces of the past is love?"

"Exactly, and there again, for several reasons. For love will rearrange past situations, wherever they are forgotten, in order to build new situations that will give causal justification for this love being poured into the present. That is where the whole magic of love lies. When you give it, you have no idea why you are doing it, because everything occurs as if you should prepare yourself to live through something that will explain why you have given it."

"You mean that the reward comes before the gift?"

"Precisely. Weren't you talking about a retroactive loop? This loop and your cycle of Love are one and the same thing."

"Which would explain why some rewards are overwhelming, and don't obey the principle of... um, how do we say it?"

"Just deserts!"

"Yes, that's it, and that's what I don't quite understand yet."

"If you want, but it's better than that. With love, there is no just deserts. There is an incredible amplification of its effects. So incredible in fact that it's better for you that it should turn into magic."

"Really? Why is that?"

"Because then happiness is easier for you to live with. That is why love turns into magic when it truly overflows. Generally, it turns into lots of other things."

"Like what?"

"Where do you think all your new ideas and creative energy come from? Love is channelled into creation!

Love is the origin of all creation, including the creation of the universe, which is why God is Love."

Chapter 20—Summary

A way of explaining all the phenomena we habitually classify under the label of irrationality: divinations, the placebo effect, miracle healing, parapsychology, astrology, etc.

New perspectives on the manner of approaching these mysteries that have come down through the centuries, whose impact we can now evaluate, by finally understanding that their foundations are not entirely devoid of truth.

The extraordinary power of the gift of self in healing, and of love in bringing harmony to all situations.

For well beyond the just—and causal—rewards of giving oneself, the person who begins or who sustains the cycle of love realizes the magic that considerably increases its positive effects.

And finally, love's expression as creation, love being the creative energy that characterizes the expression of the Mind itself.

PART V

LECTURE

XXI.

PHYSICAL FUNCTIONS OF CONSCIOUSNESS

"A flexible cylinder to model physical functions of consciousness"

This text is based on a lecture delivered at the Institut de France, in Paris, on 5 December 2012, as part of the symposium on "Birth, emergence and appearances of consciousness" organized by François Gros, Perpetual Honorary Secretary of the Academy of Sciences, François Terré, honorary doctor at law, member of the Academy of Moral and Political Sciences and Bérénice Tournafond, President of the "Being Human" Society.
It has been published in *Cosmopolitics* Vol. 2, 2013.

The author Philippe Guillemant is a Researcher at CNRS (French National Centre for scientific Research), engineer from the "Ecole Centrale Paris", specializes in the Physics of Information and Chaos. He obtained his PhD and Research Supervision habilitation from Aix-Marseille University, and has been awarded the "Cristal du CNRS" and the "International industrial vision trophy" for his innovative work in image processing and artificial intelligence.

Abstract

We consider the universe as entirely described by physical information and we propose to describe its evolution as a variation of its information, as if it were calculated by a giant computer implementing the laws of nature. However, a delicate aspect of this process is that nature contains a certain amount of indeterminism, something demonstrated by quantum mechanics. We interpret it as a lack of physical information and we show that many chaotic or dispersive macroscopic systems could also lose information. In such a context, where information about reality could be not only lacking locally but also automatically lost, that is to say turned into macroscopic quantum states, we propose that consciousness could have two functions: firstly, recording new information in the universe in the present, in the form of classical states; and secondly, configuring information about the future as probabilities of events, meaning that there is a coexistence of multiple branches of events within a multiverse. Though a choice seems to be made by consciousness in the present among various possibilities, we show that this choice is not a result of free will because it depends on an already realized immediate future, with respect to the first function of consciousness, which is purely a deterministic recording. We propose that free will could be obtained via the second function of consciousness, which is to modify probabilities in the future, by means of external information that could come via extra dimensions. We show that this mechanism for configuring future information could explain coincidences and synchronicity, involving retrocausality. We finally choose a flexible cylinder as a very simple reduced model to describe the effect of both functions of consciousness on our individual path of life and the effect of extra dimensions on the form and orientation of the cylindrical path towards the future.

Introduction

Good morning, I am very honoured to be here with you all today. I am a physics engineer conducting research at the National Center of Scientific Research (CNRS) on the subject of physics information. This has been an emerging field ever since physics began giving an objective meaning to information, something we can sum up as comparing the universe to a giant computer. This approach will allow us to improve our understanding of the fundamental role played by consciousness. It is a question of finding out what sort of relationship the universe has with consciousness, to what extent it exists outside consciousness and what role this plays in the structure of the universe. As you will have seen from the subtitle of this conference, I have already made an immediate reply to the question regarding my own point of view about consciousness: "Recording the present and configuring the future." In fact, there is no general point of view on this subject among physicists, because they tend to avoid it altogether, expressing only personal views. My own view is, I admit it, an original one: stating that consciousness records the present and configures the future may appear disturbing, so before I start, let me make things a little clearer.

What I am talking about is a recording INSIDE the universe and a configuration OF the universe. Yes, that does mean that the universe needs our consciousness as an interface for acquiring information, for here we will consider that the universe is a universe of information. But let me reassure those of you who are attached to an objective conception of reality. My viewpoint is one that sustains such a concept, that is to say, of a universe actually existing outside ourselves, a material universe where causality plays a fundamental role, except that this universe is not entirely structured. Certain pieces of information are missing, which is why those of you with spiritualist leanings will also find it interesting. I have to say, though, that it is highly likely that consciousness doesn't actually structure anything most of the time, that the universe is already structured and that everything happens as if reality was independent of us. Nevertheless, we will develop the premise that this is not always the case, or to put it another way, that the universe is only partially configured, something that quantum mechanics already shows us to be true.

Inverting received wisdom on key concepts of Time and Information
Another key concept, besides information, is fundamental for gaining a better understanding of consciousness, and this is the concept of time. In my presentation, I will be talking essentially about time and information, and I will quote three famous men: Einstein, Nietzsche and Bergson, to help you understand why it is interesting to link the three of them together:

In a letter to his friend Besso, just before his death, Einstein wrote: "People like us, who believe in physics, know that the distinction between past, present and future is only a stubbornly persistent illusion."

Such absence of distinction means that the past, present and future are simultaneous or in some way contained in the present. It begs the question whether this implies that the future might already be realized. Nietzsche wrote about this, in "Human, all too human": "Our destiny exercises its influence over us even when, as yet, we have not learned its nature: it is our future that lays down the law of our today."

The following phrase is also attributed to him, and it helps us understand: "The future influences the present just as much as the past."

Astonishing, don't you think? In any case, it has the advantage of being even clearer than Einstein. Above all, it is extremely awkward, because if we are determined simultaneously by our past and our future, it is difficult to see how we could retain any free will. Free will, therefore, would be just an illusion. Bergson rebelled against this determinist idea and wrote in "The possible and the real": "What is time for? Could it be the vehicle of creation and of choice? Doesn't the existence of time prove that there is indetermination in things?"

At first glance, this is in total contradiction to the two previous quotes. Indeed, it is hard to see how indetermination could be reconciled with double determination coming both from the future and from the past. But what is indetermination, in fact? Well, we will simply interpret it as a lack of information, which is why information, like time, will be a key concept during this lecture. And what we are going to do will end up turning our received wisdom about time and information completely on its head.

Here is a brief glimpse: often, we believe that time is something objective, because it is the variable "t" in physics equations. We will see that this might not be true, that time might just be subjective, because time and the variable "t" can be eliminated from physics equations.

As far as information is concerned, we usually think it is subjective because it is relative to the person that possesses it. We will see that, on the contrary, information is a concept in physics that is starting to be seen as objective, or in other words, information is in the process of becoming an actual measurement in physics.

Information: A New Measurement in Physics

Where does the idea come from that information might be an actual physical measurement? Strangely enough, it comes from traditional mechanics, thermodynamics to be more precise, a science quite as odd as quantum mechanics, although not many people know that. Everyone thinks that traditional mechanics is not odd because we believe that it is determinist, and yet the idea that traditional mechanics is determinist is at the root of an open-ended problem that first emerged 140 years ago, and still hasn't been resolved: we are talking about the thought experiment known as Maxwell's Demon.

Take a container full of gas with a wall dividing it into two compartments. A small opening allows for communication between the two compartments, meaning that the gas is at the same pressure on each side of the wall. But now imagine an apparatus capable of shutting the opening every time a molecule arrives from the left compartment and of opening it every time a molecule comes from the right: now a difference in pressure is created, and the apparatus can even create a vacuum on the left side. Maxwell's Demon is a metaphor for such an apparatus, which uses little or no energy at all, since all it has to do is open or close a sliding door with possibly negligible mass. This means that the Demon could manufacture mechanical energy simply by using the information it has about the position of molecules! This would mean that we had something capable of creating free energy! Which means that there is an acute risk that information could actually violate the laws of physics!

Today, we have managed to exorcise the demon by saying that information necessarily has a minimal cost in terms of energy equal to k T ln(2), where T is the temperature and k ln(2) is a quantum corresponding to 1 bit of information, k is Boltzman's constant, something in the region of 10^{-23}, a tiny number. Introducing 1 bit of information thus reduces the entropy by necessarily introducing energy. This means that entropy is quite simply the opposite of information.

When the demon acquires information, it consumes some of the energy related to the necessity of observing a molecule, which means that no energy is free; and vice versa—when the system loses information, it dissipates energy. This is called Landauer's principle, verified in experiments carried out last year, in 2012, by physicists from the Ecole Normale Supérieure in Lyon.

For all that not everybody really agrees yet about this correspondence between energy and information. For a start, we can doubt the necessity for the demon to need to observe the gas molecules in order to know their position, since according to determinism, this can be calculated. Therefore determinism has a problem with the correspondence between energy and information. Wanting to preserve it makes things very complicated, and in order to clarify the debate, we could say that there are two points of view:

Either we preserve determinism within a determinist framework by solving the problem that appears to make information dependent on the observer, the person who possesses it, so as not to have a subjective production of entropy or energy, which would be quite unthinkable! Some physicists believe that they have found a way to solve this problem by linking information to algorithmic complexity. Now this isn't my cup of tea at all, and frankly I find it quite impossible to digest, but it is still one possible way out for anyone unwilling to adhere at once to my way of talking, although I must point out that quicksands lurk in this direction.

Or we consider that information is one of the universe's genuine physical measurements but that it is limited everywhere to a certain number of bits because of the quantum of information, since the energy in a system is always finite. The problem here is that systems like this one—mixtures— eventually lose all their information as they become mixed and become indeterminist; in other words, even the universe itself no longer knows where the molecules are, just like in quantum mechanics. This is something I have verified for myself by calculation, through my work with billiards, something we will say more about later.

Classical physicists find it hard to accept this idea. And yet, all we are doing is finding the same result as in quantum mechanics. We think that observed reality isn't quantum because when we observe it, it appears to be determinist. But maybe the observer restitutes information through the act of observing. It may well be that, as a complement to what we call the decoherence process, it is the observer that redetermines reality in the end, just like in quantum mechanics.

Quantum Mechanics and Information
What happens in quantum mechanics? It too has a problem with information, because of the superposition of states and the famous observer paradox that reduces all states to one single state. You must have heard of Young's double-slit experiment in which photons are sent through one by one and we find that they pass through both slits at once, because they continue producing interference. Except, that is, when we observe them, in which case they only go through one of the slits. So as long as we don't observe them, they take every path at once. This is because the information on the path taken by the photons simply does not exist before observation. In fact, today we know that the result of observation is completely indeterminist, that is to say, it is not the result of the past.

This was proved in 1982 by Alain Aspect, in his experiment on twin particles. It proved that quantum mechanics was right and Einstein wrong when he said that God didn't play dice. In fact, it would seem that He does just that, in fact, because when you observe something, everything happens as if the universe were drawing a straw to decide what you would see, as if it were creating reality at that very moment. Einstein, though, didn't agree about this drawing of straws, and said that there had to be hidden variables, whereas Alain Aspect proved that there aren't, or at least, no local hidden variables. There is still a possibility that Einstein was actually right, but only if the hidden variables are non-local, or in other words, that they violate causality, and indeed, we will see that this is in fact the case. Having said that, the concept of "non-locality" contradicts Einstein's relativity while at the same time being fully verified today, for it is the consequence of a phenomenon we are now well acquainted with: quantum intrication.

This intrication is the second oddity of quantum mechanics. Einstein called it long-distance phantom action, that is, communication between two points in space where the signal travels faster than the speed of light, something that runs counter to the theory of relativity. In reality, there is no phantom action, and we will have to get used to the idea of intrication because it has been proved over and over again and constitutes the technology on which future quantum computers are based. Intrication is simply due to the fact that when superposed states of two particles are correlated to their origin, the correlation subsists whatever the distance between them, otherwise the mechanics of their paths would be incoherent once it had been observed, so it isn't as mysterious as all that. It is just an illusion of strangeness that comes from the fact that we cannot stop ourselves from reasoning in ordinary time, that is, from separating the moment of observation from the moment the correlation occurred.

There are other strange things in quantum mechanics, like its apparent retrocausality (or retrocausation). I say "apparent" because this is something certain people still have trouble accepting, but once again, it is because we reason in ordinary time. This is the delayed choice quantum eraser experiment that occurs with intricated photons. The result of the experiment is that everything happens as if intricated photons were capable of adjusting their past according to the future choice of a researcher in switching a detector on or off. Astonishing, isn't it? You yourself can create a part of the universe's past by looking up at a starry sky and receiving an intricated photon. This is what makes Stephen Hawking say that it is the observer who creates reality and that this creation travels back in time, which, of course, is true, as long as the universe isn't already informed about the past. This is what he calls top-down cosmology. Careful, though—Stephen Hawking remains a convinced determinist, a partisan of the theory of what are called non-local hidden variables, or variables from an unknown source. Roger Penrose does not agree with this at all, because he thinks consciousness intervenes in the process of state reduction: broadly speaking, it isn't God who's playing dice, but consciousness, in an orchestrated manner in the brain. This is his Orch'OR model, and we will come back to it later, but you can see that here, too, Penrose introduces information from an unknown source, because that is precisely what indeterminist chance is.

What we can say about quantum mechanics, then, is that in every case, it uses information that determines reality but comes from nowhere, what I call foreign information: either non-local hidden variables, or information with a non-causal influence over the present—outside the cone of light, for the connoisseurs among you—or else information resulting from an indeterminist chance event: in other words, God playing dice with wave function collapse. You have to admit that, for an exact science, this is pretty bizarre. But be careful: the main problem with all this is that the foreign information that determines what we observe is supposed to arrive in the present at the moment of observation.

Now we have a problem—because physics is in the process of eliminating the present from its equations!

Reality is not Created in the Present!

Physics has a problem with the present because of Einstein's equations and relativity. No present exists that is common to us all. If two people in the same place are both travelling, so that one of them is going much faster than the other, then this is enough to make them disagree about the present. Hardly moving at all relative to one another, but simply being very far away from each other, will also be enough to make them disagree. This is why Einstein introduced the model of the block universe, a model I show here represented by a cylinder. The circular section of the cylinder represents a disc corresponding to two dimensions of space rather than three, while the length of the cylinder represents time. The disc replaces a sphere meant to contain the entire universe. This way, the cylindrical representation includes all of space-time, from the past to the future. So we already have grounds for doubting the question of whether reality is created in the present.

Obviously this runs counter to normal intuition, for we are accustomed to think that the future does not yet exist. As far as the past is concerned, we aren't entirely sure, but we want to believe that it still exists. According to Einstein's vision, however, we see that the future is treated like the past, meaning that it is quite possible that the future already exists. Since we cannot trust the present, everything occurs as if it didn't actually exist, a fact confirmed by the theory of loop quantum gravity. This is a very powerful theory, because it reconciles relativity and quantum mechanics, which is the main problem in physics today. It eliminates time from physics equations, in other words, the variable t representing the present. And when the present disappears, we have a problem: how does the universe evolve?

Maybe it doesn't evolve at all, in which case everything, including everything we will live through, has already been created. Einstein didn't really believe this, which is why he spent his whole life searching for a theory of grand unification, but in vain. Now, what can we do to rediscover our liberty in a space where our life is fixed? I have illustrated this using the red temporal arrow representing our destiny. In four-dimensional space-time, this line cannot move. But maybe there is a solution to make it move?

Carlo Rovelli, one of the two progenitors of loop quantum gravity, thinks that the block-universe can move, but not in the present. He wrote: "We have to learn to think of the world, not in terms of something that evolves in time, but in another way." Now that is exactly what we are going to try to do, bearing in mind that Carlo Rovelli made Space-Time vibrate.

A Solution: Making Space-Time Vibrate

So here we are, I've portrayed several of these temporal lines bottom right, with a horizontal bar representing the present, except that now we know the universe cannot evolve any further in the present. The question, then, is this: if we made Space-Time vibrate, couldn't we make all the lines evolve?

Apparently not, because quantum gravity only makes Space-Time vibrate at an infinitely small quantum level, completely undetectable at our own. But we're forgetting something very important here: sensibility to initial conditions, or chaos! Using calculations, as we will see later, I have personally verified that infinitely small vibrations at a quantum level can definitely generate chaos, especially bifurcations capable of creating a considerable number of evolutionary variants at a macroscopic level.

In which case, whatever we were set to experience would already be realized, but susceptible to constant modification! Now you can imagine that the universe evolves everywhere in time at once, simultaneously in the future and the present.

Now, is that reasonable? In any case, mathematically, yes, it's possible, all you have to do is add a fifth dimension. Kaluza and Klein were the first to put forward the idea of a fifth dimension as much as a century ago, and it was something Einstein looked on favourably. It is a very tiny dimension rolled in upon itself. To imagine it, you have to tell yourself that instead of moving along a line, you keep circling round it. This is how we can regain our freedom, and it is by no means a false hope, because today's theories of grand unification also offer us this possibility. Let's take a closer look.

Grand Unification Theories Add Foreign Information (Extra Dimensions)

The main theories of grand unification are string theory and loop quantum gravity theory, both of which appear at first glance to be completely incompatible, tending to imply that one of them must be wrong. But in fact, we will see that they have a great deal in common. To start with, we only have to point out that one of them maintains an immobile space, here represented by the shape of a tree where every branch forms a cylinder of the universe, while the other one makes space vibrate. The difference between them resides in their mathematical approach, but in the end you will see that the two theories are in fact complementary. Instead of making space move, string theory introduces 6 or 7 extra dimensions to space, which in the end amounts to the same thing. On the one hand, you have an elementary particle, the quark, for example, vibrating in the extra dimensions of space. On the other hand, you have the same particle also vibrating, because it is space itself that is vibrating. As a result, the theory that makes space vibrate has no need to add spatial dimensions. However, it doesn't have enough dimensions to fully describe reality. It merely describes a stochastic, probabilistic reality. When the quantum fluctuations of space are transformed into real observations, then there is always a reduction of state into a single reality that is not described. Because of this, the future as described by loop quantum gravity is one that potentially contains multiple possibilities, just like the future as described by string theory, what we call the multiverse. We don't see it as quantum gravity because it is a dynamic approach that doesn't describe everything, whereas in the case of the multiverse of string theory we are dealing with a static approach that describes everything, but contains too many degrees of freedom.

Now, to marry these two theories, all we have to do is observe that the quantum fluctuations of the first may very well be responsible for the changes from one branch to another in the second.

The problem is that Stephen Hawking said nothing about all that. He said that we only live on one branch of the multiverse, but the thing is, he has no proof. In fact, there are 10 to the power of 500 possibilities of varying the vibrations of strings, and living on one single branch of the multiverse would mean that this vibration was always the same: there is no reason here, it is quite arbitrary. If you look at loop theory, quantum fluctuations can make us change futures in a completely random manner, so you see that chance can make us change branches. So if you reconcile the two theories, you have to reject Hawking's idea that we only live on one branch of the multiverse even though there are myriad others. And it is better that way, because the consequence of Hawking's theory is that it forces us to think that we all have billions of conscious lives in parallel universes. Personally, I cannot imagine that. There is a much simpler solution. All you have to do is confront those famous sources of foreign information that the other theories put in, whether they be hidden variables, chance or extra dimensions. Alain Connes, our own great mathematician, also adds spatial dimensions to space to reconcile the two theories of grand unification: yet again, additional information from foreign sources, as if, at every point in space, physics needs complementary information in order to predict the course of events, although no one yet understands where this information comes from!

So now we can ask ourselves this: why don't we use this sort of complementary information or extra dimensions in classical mechanics, at our own level?

Classical Mechanics Do Not Work Well in 3D

In fact, this sort of additional information would be very useful in classical mechanics, because if we do a little exploring we come to see that mechanics actually work very badly in three dimensions. Or at least it does if we start from the point of view that information holds real physical meaning, implying that the entire universe has to be quantified. We encounter this quantification in loop quantum gravity that requires the universe to be discrete or of granular structure: this would imply the existence of indivisible grains of space, but also of energy, impulse, time, etc. and in the end the universe would be like the picture on a television, not actually continuous but made up of pixels. Or if you prefer, the universe would contain an enormous but limited quantity of gigabytes, like some enormous computer.

All this has consequences at a macroscopic level because the indeterminism that comes from quantification can then rapidly spread to the macroscopic level, which is what I am working on.

The first consequence of a granular space structure is that classical mechanics become indeterminist in three dimensions, as we have already seen with Maxwell's demon. This is what I have been able to demonstrate with calculations using billiards[7]. Here is an animation that illustrates how, after a certain amount of time in three dimensions, mechanics cease functioning. I take two superposed billiards with the same initial conditions, except for one infinitesimally small difference between them, in this case a difference of 10^{-15}, and look how quickly it happens: as you can see, the two billiards swiftly diverge. Mechanics don't know where to go anymore. This is because it has lost practically all its information concerning the balls. I have been able to show that this happens even if the initial conditions differ by one mere grain of space—a non-existent, meaningless value. In that case,

---

7. See http://www.youtube.com/watch?v=uc5NTFrqlPU

using 500 balls, it only takes about thirty collisions per ball for mechanics to stop functioning, and when you increase the number of balls, it takes even fewer collisions, actually tending towards zero the nearer one gets to infinity.

But that's not all, because I have also confirmed that even if space does not have a granular structure and can thus be as precise as we like, we still have an enormous problem with information, this being that when you calculate the trajectories of all the balls for as long as mechanics functions, you use a certain amount of memory to stock the entire evolution you have calculated. But after a certain number of balls, this quantity of memory becomes inferior to the memory required for the initial conditions. For a mechanical physicist, this represents something utterly aberrant: the fact that his model uses more information than it actually calculates. I have called this the paradox of classical information, or the demon of determinism. This means that as soon as we admit that information has a physical meaning, mechanics won't function in three-dimensional space, or at least not for very long, even if that space is continuous.

This is all very awkward. To restore determinism—something we have to do because we inhabit a single reality—I have therefore tried to add foreign information to allow us to choose the direction of events. For the moment, though, I'm having a little trouble with the technical choices I need to make to get it right, and spending more time on the presentation of what I have explained to you.

There is something else, however: the smaller the balls are, the sooner indeterminism sets in on the billiard game, and in any other interactive system, even if we are not dealing with elastic collisions. Now this might well explain the quantum behaviour of particles at a lower level. For my part, the more I study mechanics, the less I see a difference between the classical and the quantum world, since indeterminism, observer interaction and even intrication are, to my mind, imposed by both kinds of physics. Personally, I have good reason to think that there is no frontier between the two, but I won't be insisting on this point, as I haven't yet established sufficient confirmation for my premise.

What we need to keep in mind here is that, yet again, we see that in order to calculate the future, it is indispensable to introduce foreign information to physics—except, of course, if we leave it to chance, by calculating the most probable future. The problem is that we're dealing with classical mechanics, something we don't often suspect of odd behaviour; yet the only way to introduce such information is to add at least one dimension to space, making five in all. This is nothing new, it's an idea we're starting to get used to, and something interesting happens every time someone tries to add dimensions, which is that the dimensions are always rotational. With billiards, it is a case of curving the trajectories so as to make them pass through all the possible paths they would otherwise ignore. It is as though we needed to twist space in different ways, and indeed, such torsion can easily be formalized by a vector that wraps itself around a point or a line. So now we will simplify it all by pretending that there is just one, rolled-up dimension.

What Happens When We Switch From One Destiny to Another?

Right, now we can get down to things that are closer to us and to our own lives: we are going to add a fifth dimension to the cylindrical space-time I showed you just now, a rolled-up and above all macroscopic dimensions, so that we can visualize it more easily. This dimension is going to let space-time change shape. But beware—this change is not because space-time itself is moving, for while this is true, it only moves an infinitely small amount. It is because whatever is inside may change considerably at a macroscopic level, as I explained: something quite negligible can provoke huge effects, bifurcations, new scenarios. Keep this image in mind: a disk or a slice of a cylinder represents two dimensions instead of three. So we can include the whole of the universe inside a sphere, here replaced by a slice of a cylinder. Now we can shrink this cylinder into a flexible tube, and we can even represent part of this universe, our own human life, for example, by shrinking it further into a piece of string whose length represents time.

Now let us take this rolled-up dimension and make it work: to do so, we have to twist the piece of string by making it rotate around itself in the present, and look what happens: the end of the string doesn't really change position in the present, but its position in the future changes dramatically. What we have to understand is that when we add in a dimension and then move, even just a tiny distance along this new dimension, the whole of space-time changes all at once. This is why twisting the tube or the piece of string is a good illustration for what happens, because we can see quite clearly that this rotation will alter everything in both space and time.

Using this, we can agree with the space-time of loop quantum gravity, which doesn't evolve in the present but simultaneously everywhere in time, due to quantum fluctuations. We can also agree with string theory, because by making the piece of string twist, we end up reconstructing all the possible branches of the multiverse, until the whole thing begins to resemble a tree, what I call the Tree of Life in my book (*The Road of Time*). The universe has its own tree of life, and each of us has his own, too. The tree is a static representation wherein we can see all possibilities at a glance. Here, I have chosen to show you a tree with cylindrical branches: they are like tubes. This is a vision painted by Robert Venoza.

Now, let us look carefully at what happens when I twist the tube or piece of string and suddenly change the future, something that can happen at a bifurcation, that is, when the cause is infinitesimally small. For example, if you take a route you don't usually take to go to a regular meeting, you may make the chance encounter of your lifetime, and your life will suddenly change. However, we can't really say exactly when your life changed, because it might have been long before that, if for example one day in the past, one of your friends told you you should shake up your habits a little. This might have suddenly made you become aware of something, so that when you got the chance to act on it, you automatically chose the bifurcation. Meaning that your life didn't change the day you made the encounter, but well before, the day that becoming aware took place. So let us look at what actually happened that day.

Retro-Causality Is Inseparable From Causality in an Evolving Block Universe

I have illustrated how your life can change in two ways. Top left is change using a tube, and top right is a simple graph with your future on the X axis and a foreign parameter of realization on the Y axis. The change might not have actually occurred, for example if you were too self-absorbed. In red is where it does occur, and you can see what happens: your future changes, but at that point it becomes impossible that the past immediately preceding the moment of the encounter should not change, even if only a tiny bit, because otherwise it would provoke a breach in the space-time continuum. For we cannot act upon one point in time without simultaneously acting on what comes after AND what comes before. Quite simply in order to respect causality.

Now let us take the model of the tube again, and imagine that it is an invisible tunnel along which each of us travels, automatically guided by illusory free will. The only way to retrieve any authentic free will would be for our thoughts to be able to generate foreign information that would displace our tunnel in the fifth dimension. But also imagine that, as it moves, our tunnel starts to topple over and has to steady itself on something and adjust to the terrain, meaning that inevitably, to maintain causality, a certain retrocausality comes into play to adapt our present to this new future. It is unavoidable, because otherwise a breach opens in the space-time continuum.

Elsewhere, Thibault Damour gives us confirmation that relativity allows for retrocausality and that time is perfectly reversible. The only enemy of retrocausality, in fact, is irreversibility, which spends all its time trying to pry the two twins apart. Irreversibility is in trouble, however, especially if we reason in terms of information, because the information present at the moment of the Big Bang, especially in a discrete space, obviously has to be infinitely less than the information present today, with all its living systems. And so, since information is the opposite of entropy, this would mean that entropy has decreased, something that goes against the second principle dictated to us by irreversibility. I believe that this is because the second principle has forgotten the creative effect of foreign information on the future. Foreign information guides the future, that's about the sum of it, and Darwinism needs to be revised. Now, I would like to make it clear that in fact, the question of irreversibility is redundant in indeterminist space-time, or shall we say a space-time determined in multiple ways by foreign variables. This arises from the fact that mechanics does not describe a single future or a single past, but multiple futures and pasts. And the question of irreversibility does not arise in the presence of multiple pasts.

To summarize on the subject of retrocausality, I would like to point out that more and more publications are coming out today on the subject of the influence of the present on the past and especially of the future on the present, including articles in the review Nature. I would also like to note that retrocausality in five dimensions has no need for tachyons, or in other words, particles able to travel faster than the speed of light, and lastly I will point out that Bayes's famous equation, emblazoned last month across the cover of Science et Vie, which practically presented it like some magic formula, is actually completely trivial in the presence of retrocausality.

Now we will see that retrocausality has the advantage of explaining certain strange phenomena, like coincidences and synchronicities.

The Mechanics of Coincidences and Synchronicities

Synchronicity—now here's a phenomenon that might well validate the possibility that our thoughts have some connexion with a fifth dimension, because the foreign information they introduce is the only thing able to explain phenomena that defy causality. We call this Acausality, putting a capital A in front of "causality". This concept was invoked by Jung to explain synchronicity in his correspondence with Nobel Prize-winning physicist Pauli, the father of the concept of Acausality. Synchronicity is a coincidence that is loaded with meaning for the person experiencing it. It is this meaning that expresses the fact that our thoughts are part of the phenomenon. Having said that, most of the time we experience mere coincidences, phenomena we often group into different denominations: the "Law of series", "it's a small world", "Murphy's Law", the "Pauli Effect", the "Demo Effect", not forgetting numerical series and above all, CHANCE...

We can understand coincidences using retrocausality, which has a tendency to create synchronous order in the reverse sense of time, because mechanics tend to create disorder in the normal sense. For that to happen, however, the future has to be restructured, meaning that it has to be in the process of changing, which is why coincidences are rare. It is easier, then, to understand the phenomenon by saying that when the future changes, it may be placed in a situation where it has to find new causes, if, for example, the former cause is made more improbable. But we're not really sure how this occurs, which is why synchronicities are the more interesting phenomenon of the two. As far as synchronicities are concerned, we know that they interact with our thoughts and emotions, that they contain a meaning that can make us responsible for changes in the probability of different causes.

Synchronicities can easily be explained if we consider that our intentions can cause effects in the future that become the future causes of an effect in the present. This is an idea also developed by Jacques Vallée, who declared that it will represent a future current in physics, the physics of information, in its conference at TEDx Brussels (2012). In fact, he is the one who made me explore this avenue, more precisely the problematic of Maxwell's demon, and I have to say that I was surprised to find grounds here to confirm the macroscopic indeterminism which lies at the base of my theory of double causality. Quite simply, this means that we could be living in a world of information that is only partially configured. And it is precisely this that allows us to open the field of possibilities, and that I myself voluntarily tested in order to verify my theory. I explain it in my book *The Road of Time*. I won't go into it any further now, because it would take too long. All I will say is that provoking synchronicities requires an alertness of the mind, a letting go, a genuine freedom from conditioning and an openness to changes in one's life. All of this can be summed up in one very simple phrase: everything that isn't determined by the past is determined by the future. Remember my quote from Nietzsche.

I would also like to draw attention to the synchronicity theory of François Martin, a quantum physics theoretician at CNRS. His theory on the quantum psyche brings into play a quantum intrication between two events that may be separated not only in space but also in time, and I think that this atemporal intrication is certainly a key to gaining a better understanding of what I call foreign information. Whatever the case, we will see that intrication or extra dimensions are, in the end, two ways of introducing the same thing, in this case atemporal information.

The Flexible Cylinder: a Model for Consciousness?

Now I am coming to the main purpose of this lecture, which is to find out what consciousness function could be, and whether we can approach it using a model. I think that when we possess a simple geometric shape that allows us to illustrate something as enigmatic and complex as space, time and extra dimensions, then we should keep it and use it properly.

So I will represent our brain by means of the grey circle inside the tunnel we are each travelling through, unaware that we are being guided, and I will represent our consciousness by means of a pocket lamp that lights our path. I make a distinction between consciousness and brain, because, on the one hand, as the conceiver of artificial visual brains, I have absolutely no faith in consciousness magically emerging from complexity, and on the other hand, we have dire need of a serious interface to reduce superposed quantum states, whether they are in the brain or elsewhere in our environment. Our universe of information needs to be observed in order to acquire new material information from all the possibilities available, and it seems to me to be right, as Roger Penrose suggests in his model Orch'OR, that consciousness should play this role.

However, to my mind, what we are talking about here is a passive role that might well be determinist, if in fact consciousness can forget that it is an interface between the brain and a source of foreign information. The ego, of course, is what would tend to make it forget that, something I have illustrated using a mirror for where consciousness is passive. I have also given an illustration below of a model of active consciousness, this time with a source of foreign information, able to act in the fifth dimension by being connected to the brain via consciousness.

As you will have realized, I identify this source of information with what we call the Mind. When this source is used, that is to say when there is no ego in the middle acting as a mirror, then information can be transferred from the mind to the brain by consciousness, leading to a change in the brain, characterized by a sense of awareness that instantly alters the future. There is no phantom action, therefore, and everything does indeed transit via the brain. What I am putting forward here, then, is nothing other than a rather materialistic model of consciousness and of the mind, which respects the laws of mechanics and even restores a certain amount of determinism.

To sum up, then, I can identify two functions of consciousness. The first is a passive function. Because it is related to observation in the present and because the universe needs to be observed in order to acquire information, I propose that the first function of consciousness be one of recording the universe.

The second function of consciousness is an active function for configuring the future. It consists of preparing it, informing it via the brain with the help of information from the mind, as long, of course, as this information can reach it. The future is always partially configured, and it is only as it gets closer to the present that the configuration is gradually finalized until it reaches maximum density closest to the present. Thus passive consciousness eventually realizes the possibility that has become the most dense, that is to say the most probable, by recording it. In my opinion, therefore, there is no foreign information coming out of nowhere and entering the present, because the immediate future is already prepared. The non-local hidden variables are quite simply contained in that immediate future. The action of mind configures reality in a more distant future. However, I think I am only setting out a vague sketch here. This sketch leads me to suggest a link between consciousness and gravity, because all in all, information from a fifth dimension plays the same role as quantum gravity by moving space, but we have to be careful about this. Whatever the case, you will have noticed that this rough model is at once very materialistic, while at the same time being compatible with a certain kind of spirituality.

Schrödinger's Cat
Now we are better equipped to try to interpret what is happening inside that famous box holding a cat in danger of dying if the apparatus shut inside the box with it suddenly releases a deadly poison. Within the time allotted to the experiment, this apparatus has a one-in-two chance of being activated by a

radioactive particle. The problem is that this emission is subject to the laws of quantum mechanics and as a result, the moment it occurs is determinist. But that's not all. Most importantly, there is a possible superposition of states between the alternative where this emission takes place and the alternative that it doesn't, or in other words, the cat is dead and alive at the same time!

Physicists have argued long and hard over this problem, because a process called decoherence prevents such a superposition of states from existing at a macroscopic level. However, this is not entirely true, because, first of all, we have already observed superposition phenomena at a quasi-macroscopic level, such as in interference experiments for example, using giant fullerenes, or even superconductors or within the process of photosynthesis. Next, decoherence theory does not properly explain the final phenomenon of state reduction where we cannot in the end manage without an observer. It would seem that this reduction happens naturally, in the absence of any observer, but that it is a question of time, that is to say, there is a very brief lapse of time during which the superposed state exists. So the cat is certainly already either dead or alive when we open the box, but the thought experiment remains valid in theory, because we are allowed to consider that it is indeed dead and alive at the same time, for a certain amount of time.

Now let us suppose that the superposed state actually exists. The real question is whether there is any difference if the cat is conscious or not, or rather, whether it is a real cat or a robot cat. If it's a robot cat, then it is an authentic dead-AND-alive Schrödinger's cat, but if it's a real cat, then it is in some way designed by the universe to provide it with information via its consciousness, in which case it is either dead OR alive. This is where the subtlety lies. And if it's a highly talented human being, capable of provoking its own chance, then we could even imagine that it might be able to stay alive, although we're moving rather into fantasy here, obviously. Whatever the truth, it raises the problem of what happens when the cat is asleep, and that's a question I'm not even going to attempt to find an answer for...

Conclusions

To conclude, I think that consciousness has a dual function that can reconcile both the materialistic and spiritualist approach. The materialist approach consists of recording the present and does not reject the neurophysiological determinism affecting the conscience, while the spiritualist approach is concerned with the future and consists of saying that via the mind, consciousness can have an impact on the future. This way, we can reconcile Bergson, Einstein and Nietzsche. Double determinism does indeed exist, but time remains a vehicle for creation and choice nevertheless, via the mind.

Lastly, what I call the theory of double causality, which is quite simply the name I give this model of consciousness, allows us to explain coincidences, synchronicities and even the differences between them. Now, there may be other strange phenomena that could be explained by this theory, but I have preferred to concentrate here on synchronicity. If I could sum up this theory, it would be to recall that it is founded on the macroscopic indeterminism that allows for a possible evolution of the universe outside of time, and that this evolution also appears necessary when we try to reconcile the theories of grand unification.

Perspectives

As I draw to a close, here is something I haven't mentioned in my lecture because it hasn't yet been published or confirmed, but I hope, and I have good reason to believe, that the theory of double causality could offer an explanation for dark matter on several levels: the fact that it predicts how the radiation of a star or other matter could suddenly disappear, something that has already been observed; the fact that the matter whose radiation has disappeared should occur all the more frequently the further away it is; and finally that this selfsame radiation might reappear at any moment. All this could be a direct consequence of the fact that the universe's past might evolve, obviously infinitely less rapidly than the future, because of the enormous quantity of traces the past leaves in the present, but even so, it could evolve. The consequence of this would be that former pasts could temporarily block the light... but I won't go any further, because we find it very hard to grasp the concept of these stories about evolving pasts.

Finally, I believe that none of this is very significant compared to what, for me, is the most important perspective opened up by this model. These are things I have experimented with personally, things I have spent a long time thinking about, until I can go so far as to say that I have become deeply convinced of them. I have come to understand that moral and spiritual values such as detachment, giving, authenticity, confidence, faith and intuition are values that are a natural result of the most appropriate behaviour pattern, once we come to understand the nature of time and the information this theory suggests.

Thank you for listening.

CONCLUSION

When I started writing this book, I had only an intuitive idea about the mechanism of synchronicities, and I was a long way from imagining that by working on this intuition, by analysing it and structuring it, using teaching from modern physics, I would be led to spin a long, long thread that would lend substance to concepts as mysterious as the mind and soul, and, more astonishing yet, give a physical functionality to Love, as a fundamental substance for carrying the effects of second causality.

Were these just wild imaginings on my part? This is not the conclusion I arrived at, conscious meanwhile that I had reached limits beyond my capacities as a mere mortal. And so I took the risk of free thinking.

I was far from suspecting that this theory, finally emerging unaided from the presupposition of our free will and the omnipresence of the future, would thus provide a way to explain many other things apart from coincidences: instinct, intuition, divination, the placebo effect, even alternative medicine and the surprising effects of oriental disciplines, aiming, as I see it, to augment our receptiveness to second causality, using the practice of emptying one's inner thoughts.

All I am doing here is identifying new threads or guiding principles, but obviously I am not going to "spin" them myself, for this would be a phenomenal task. Even if I think that I have reached my goal concerning coincidences, in this precise area alone there is still considerable work to be done, if only merely to bring the subject of time into the field of scientific experiment.

The subject of time has been approached in a remarkable way in several literary works by the French physicist Etienne Klein (12) (13) (14), to whom I pay homage, because he is the one who opened the breach I crept into. To my knowledge, he is the first physicist to have made a crack in the wall of irreversibility, by pointing straight at the very place where something wasn't quite right in physics, and starting to tap away at it. One might lament the fact that, having seen him create or design this crack, I came along after him with a great sledgehammer, but anyone who is familiar with the crack might also point out that all I did was break down a door that was already open.

I can already hear my more sympathetic fellow physicists telling me: "You shouldn't have said such things!" But it is already a delicate matter vulgarizing a difficult theory that everyone agrees on, let alone one that is controversial because one was bold enough to answer "yes" to the following provocation: "Do you claim that a tornado could rebuild a house?"

If, before that crack in the wall of irreversibility, we couldn't really see, or didn't want to, that it was an artificial construction, it was because the considerable accumulation of tried and tested results and the immense technological fallout of statistical physics had encountered quite a dazzling success!

However, there is some kind of imbalance within this school of physics, between its experimental and technological potential on the one hand, and its basic equations on the other, which are founded on chance, that is, on statistics and probabilities: this is an empirical edifice, and one lacking a fundamental vision of elegance. Since, in my opinion, this lack of elegance points to an incompleteness that reveals an essential omission, it has, if my diagnosis is correct, contributed to the lengthy perpetuation of an illusion in our vision of the world: that of causality!

For chance is causality's bosom buddy, the one that urged it to keep her sister shut up in the basement, when many clues were leading us to believe that causality did in fact have a sister.

That being said, keeping everything in proportion and putting caution to one side, it is science itself, not religion, that today invites us to ascertain that chance merely bears witness to the fact that this place still remains vacant, by way of the many achievements that have contributed to reevaluating man's place in the universe, notably via the anthropic principle* (27) and the paradox of the observer in quantum mechanics. But if we only dare to sit there ourselves, we realize that as humans, we are co-creators of our reality, and co-creators of the universe itself.

So wouldn't it be better for us, human beings, to progress while integrating this new knowledge, instead of spending all our time denying or ignoring it?

We have to accept the obvious: all I am doing here is hammering home a message that has already got through to non-scientific minds, and the scientific system—assuming it to be a unitary body—doesn't take in hand the new paradigm gradually being integrated by humanity without its help, then more and more science will become reduced to operating exclusively in the realm of technology; and what will emerge will be purely and simply a new system of beliefs seeking to impose on us ideas and precepts about life that we understand no better than we did before!

And when we are incited to "Love one another!" we will again reply: "yes, of course, we're all agreed on that, but we'd really like to understand why?"

Let us admit that we have absolutely no need of science to help us love our neighbour, and let us see the consequences of science thus withdrawing from everything that affects human beings. Let us simply take the example of clairvoyance: science tells us that we shouldn't believe in clairvoyance because it's just a load of nonsense that comes from charlatans.

What does double causality have to say about it?

Well, double causality theory tells us that we shouldn't believe what clairvoyants say, because all such visions are apt to change all the time. Isn't that simpler, and a little more respectful, after all?

I could go on to quote many other examples in the same vein. In fact, science would only have to start taking an interest in double causality for us finally to rid ourselves of these endless, sterile debates that oppose rationalist and intuitive thinkers, the partisans of "common sense" on the one hand and faith on the other, and so on.

Just because we are rational doesn't mean to say that we are not intuitive, and vice versa. We have to stop perpetuating this opposition between adepts of first causality and adepts of the second.

Both causalities are complementary!

We are well acquainted with the first. The principal interest of second causality is that it offers us a new vision of the world in which the human mind takes up its proper place, no longer limited to a conscience that would appear to be a by-product of the neuronal processes at work in our physical brain.

What follows is spiritual elevation, greater responsibility, increased moral sensitivity and self-control. For if second causality is based on the fact that our thoughts and particularly our intentions have active results, very fortunately they only do so in exchange for fundamental personal qualities that require self-sacrifice and the setting aside of the ego. That is why we have reason to fear the results of increased sensitivity to second causality. We simply have to beware of manipulators who try to make us mistake empty causal vessels for non-causal lanterns.

Self-sacrifice which appears in the form of Love has the ability to reshape the amplitudes of probabilities attributed to our different future potentials, and this has non-causal consequences that translate into the sometimes impressive amplification of the probabilities of finding paths that will realize our intentions.

We find it hard to get used to this idea of a future that we could modify all the time, because we believe that beyond our present, nothing exists. Whereas this is a false belief, and Einstein was the first to prove it to us.

The omnipresence of our future does, however, have one very interesting consequence, for even though our free will is apparently increased to an excessive degree by second causality, it is nevertheless the case that most of the time, our future is already determined, that is, fully mapped out. We move forward in time as if we were inside a large tube placed along the whole of our future. We think of our actions as being entirely free, we think ourselves capable of controlling them at any moment, but in reality, nothing changes, everything is already there. The tourist travelling in a coach or on a barge feels free and may even be very happy, because he can move about on his vehicle, take an interest in everything he wants around him, change what he has for breakfast, but he will still go to exactly the same place as all the others, and he won't change anything about his life that has already been memorized by the universe.

This tourist will only achieve a true act of free will if he decides to leave his mode of transport and get by under his own steam. But at that moment, the tube he is in merely takes up a new position, and generally not for the whole stretch of his future. Most of the time, he will merely have shifted it a little, for a single day.

This is how, as a general rule, we change absolutely nothing in the structure of the universe, past or future, and when we do change something, our tube merely repositions itself, and the universe simply restabilizes its structure once again, past and future. Here is the completely paradoxical consequence of second causality: most of the time, we find ourselves confronted with the illusion of our free will!

In the end, though, isn't it better like that, rather than supposing that, starting with a present that doesn't even exist, the universe endlessly recreates itself as if it were waiting for us to find out what it should do? This latter vision of a universe "waiting" for mankind is surely exaggeratedly anthropocentric, or even egocentric. Isn't it nonsense to consider that the universe needs our existence in the present in order to know how it should evolve, as if its desire were merely that of satisfying all our immediate orders?

If general relativity destroyed the notion of the present, the same definitely goes for quantum mechanics! For with decoherence theory, it teaches us that we have no responsibility at all for the contents of the choices made by the universe at the present moment of our observations, for it is our surroundings that prepare them and not us. In any case, science has always encouraged us to believe that the universe does not need us in order to exist! Without us, it could still exist in all its past and future possibilities, and there too, there would be no notion of the present.

So it is clear that physics is so embarrassed by the present that it tends even to expel us from it! Which is why, trying so hard to do without it, it ends up by managing without us completely and finding its most coherent expression in the theory of parallel universes, without giving us any reason for accepting all this waste of space other than telling us that we are nothing after all.

Whereas if, on the contrary, we are important for something, what could it be if not for imposing the existence of a single, unique reality out of all these universes?

What reality, though, could we give preference to, since physics forbids us to choose it in the present?

If we truly are endowed with free will, how do we exert it if the universe chooses the present instead of us?

Indeed, we cannot make this choice. We cannot control our future in the present. All we can do is observe and make what we observe enter reality. Then everything depends on what we are going to be able to observe, and so everything depends on how alert and attentive we are. For if we are sleepwalkers inside a tube that doesn't move, then it is obvious that only chance might actually move the tube!

But even this is not the case, for even chance is impotent!

Chance does not "move" anything at all, seeing as it isn't real! If it was in fact real, then it would be deployed in all its possible versions. Whereas by definition, what is real is unique. True chance exists but is not real, because it is destroyed by its entry into reality, by the observation that turns it into a false chance: an event that is not random, but whose causes we do not yet know because they are hidden from us, given that they come from the future!

There is, however, a healthy need for chance, when observed as such, in terms of a joker to be played when we don't know how to interpret it! And I recommend using such a card without moderation!

We should not, however, tumble into the opposite excess, like those who play the stock market, for example, using the laws of causality and especially its close friend chance to an exaggerated extent, lining their own pockets while at the same time creating false value, while our industries are drained of their resources.

It is high time, then, in view of the current financial and economic crises, that we asked ourselves whether, by accepting chance as a false brother to causality, we may well have fallen victim to an enormous deception!

This collusion between chance and causality that prevents us from seeing second causality emerge is also at the origins of other deceptions, such as natural selection. It is not a question of calling into doubt this principle at work in nature, but of denouncing its exclusive status that strives to be the only explanation for everything. Darwin was only partly right. Nature employs principles other than the law of the survival of the fittest or most adapted when making species evolve, because contrary to what they teach us in school, intentions are indeed at work in nature, and in contrast to the result of chance, they are extremely generous.

These are far from being the only errors to which causality has led us, and they are often very serious. One only has to observe the world today to notice that those responsible for its degradation—and even for its disintegration—are economic systems built on principles inherited from exclusive causality (unbridled competition, dialectic materialism), whatever their political leanings. Including second causality in the reflexions that must lead us to a new society, then, is long overdue.

Second causality obviously implies a higher level of implication, investment, alertness or "vibration" from our being. It invites us not to understand but rather to live, and more precisely to be and to love. The existential vacuums of our consumerist societies are a result of the fact that they are founded on "not being", on ignorance of the true meaning of our life, the meaning that consists of creating new harmony by exploiting our best capacities for co-creating the universe, and on a more modest level, our planetary environment. Because in the end, we create what we are and what we love. Our ignorance of being has incited us to neglect our environment, and has done so even though we have evolved from ancient tribes that were extremely sensitive to and respectful of it. They intuitively perceived this meaning of their life, although they didn't understand it, through dialogue with nature. I pay tribute here to our animist* ancestors, whose visions we often condescendingly consider as foolishness, when what we are doing ourselves is analysing their psyche in the most foolish manner possible. We have some serious lessons to learn.

How, though, do we find the right balance in our thinking? For there is obviously no question of actually abandoning causality!

If first causality teaches us to worry only about the present, second causality can become very effective when it stops being interfered with by a mind that reasons obsessively and repeatedly, endlessly projecting future scenarios. Such scenarios serve only to call into question our intentions and cancel out their effects by simple reason of ceaselessly trying to adapt them to realities. When, on the contrary, intention escapes from the present by giving the impression of falling into oblivion, then it can quite serenely invest the future and begin to take effect: let your love of life flow out gently into space-time towards the present in order to make your dearest wishes happen, though not without a little preparation, of course.

If first causality is prepared to delegate the future to second causality, it too can in turn become extremely effective in the present, as it ought to, in order to make us focus on the actions that will allow us to prepare the realization of our projects. The mere fact of not having to think about the future because it is already working for us, actually liberates that large, overcrowded part of our brain that was only inhibiting our actions in the present.

Thus both causalities are entirely complementary and could lift up mountains if only they knew how to act in unison. And today, we need more than ever before to lift up mountains in order to heal our planet and rebuild our system of economic exchange. This cannot happen without balance, simply by praying, for example. Even if prayer is useful, we cannot wait for heaven to do it all. This, however, is an attitude that many people we could put into the "New Age" category seem to have, people who generally distinguish themselves by their excessively lax and confident behaviour. Even though we do need to recognize certain advantages to "rising above it" when faced with certain situations, we must also decry here the excessive "letting go" whose consequences can be just as serious as that sliding into the opposite excess from "causal sources" that is the domain of the manipulation.

Manipulation is far from being the only excess of causality. We know, often because the cinema has introduced us to it, just how insidious the activities of secret services can be in the name of the "good cause", so much so that the question of the quality of the cause itself ends up disappearing to the benefit of the strongest power. Which is how the principal excess of simple causality is generally translated by the disappearance of all morality, in the name of organized intelligence, and at best in the name of a collective interest that only manages to impose itself to the detriment of another collective interest.

It is fairly easy to extract the excesses or abuses of the two types of causality, because the balance is a fragile one.

As far as first causality is concerned, we have already identified its main weaknesses, ego and fear. To compensate for these weaknesses, first causality has instituted power and manipulation. Power, the most positive manifestation of the ego, and manipulation, the most effective antidote to fear. Power represents the best way of valorizing the ego of the person who possesses it, and manipulation is, in the absence of power, the best arm for deflecting fear of other people's power.

Thus, shutting up her sister in the basement leads causality to manufacture these by-products, ego, fear, power, manipulation and immorality.

As far as second causality is concerned, we can affirm without risk of error that here, "drift" can be seen in the form of laziness and self-indulgence, on the one hand, and sectarianism or excessive devotion on the other. They result from an unbalanced application of the two main principles of second causality that our religions teach us in the shape of: "Ask and you will receive," and "God helps those who help themselves". As with causality, a balanced application of these principles requires us to work on self-improvement, outside and in. When the balance is destroyed, and we make only an inner effort by asking but not helping ourselves enough, we stumble into the first trap, self-indulgence. When, on the contrary, our excess weighs on the side of a predominating quality of action compared to the quality on inner investment, then we stumble into the trap of disciplines and precepts of life that are dictated to us by sects or "invasive" religions that claim to know the Way of God, something they use to enslave us.

Thus we can see that it is not easy to achieve a balance, but that in any case, it consists of always staying on the central path, avoiding extremes. Looking at this way, it is essential to listen to the lessons of Buddhism, even though this is a religion that appears to favour second causality.

Do we live to be happy? I don't believe so, because the meaning of our life is found in what we have to give, and not what we have to receive. The greatest happiness resides in self-sacrifice, for this has amplifying virtues. We can't see these virtues, because they require a little time. Time for everything we give to impregnate the ground, make the brooks swell with water and gently flow and infiltrate the soil until the magic can happen.

Then chance and happiness start to rain down, and this amplifies our gifts in an infectious manner. If the amplification happens even when we are alone, then it is even more important when we are part of a group, because then we don't just receive from ourselves, but also from others.

Here is the secret of the magic of time: Love, that can be created and self-amplified by each human being, making him into a small part of God.

Perhaps the power of second causality could be known by this very Love, by our sensitivity to its magic, revealed by the inner light emanating from each and every one of us, shining ever brighter, the higher our spiritual elevation becomes?

THANKS

My thanks to Alix de Montal
for intervening
when my Angel
asked her to do it.

GLOSSARY

Acausality: the principle whereby some events, that we cannot explain using the law of causality, are linked through meaning and resemblance.

Actualization: the process of a future potential entering present reality. Actualization is triggered by observation and depends on a choice made by free will or by conditioning. It transforms a possible existence into reality, or in other words it chooses a unique reality to be lived in the present, from a set of potential realities that already exist beyond the present.

Animism: a philosophy that attributes a soul to natural phenomena and objects. Animism is the only philosophy intrinsically to understand the possibility of a dialogue between mind and nature, a dialogue permitted by double causality.

Anthropic (principle): the observation that the universe has been carefully designed for the emergence of life and of conscience.

Anthropocentrism: the philosophy which makes man the centre and the ultimate purpose of the world.

Bifurcation: a sudden modification in the behaviour of a dynamic system, making it pass into one or other of two very distinct states, sometimes in a totally unpredictable and indeterminist manner. More generally, used to describe a crossroads between two possibilities of evolution that seem likely to depend on "free" choice.

Big Bang: a theory that the universe has been in a state of expansion and cooling down for about fifteen billion years.

Butterfly effect: used to describe a phenomenon so sensitive to initial conditions that it can translate into a minuscule cause being able to produce gigantic effects, like the beating of a butterfly's wings creating a storm.
Chance: an unpredictable turn of events, or the unpredictable evolution of a physical system. There are several types of hazards, and among them we

can quote those caused by dispersion (see dispersive hazard) and those due to chance. Hazards are interesting in double causality insofar as they are indeterminist.

Causality: a fundamental principle whereby causes come before effects.

Causality (second): causality in the reverse sense of time that relies on inverted determinism. Second causality is therefore determinist retrocausality, an interesting concept because it solves the problems posed by retrocausality, the reality of which is debated. Another reason for differentiating second causality and retrocausality is that the latter supposes that the past "precedes" the future, something that is not the case of second causality, which supposes that they are simultaneous.

Chance: used to describe an event whose causes are unknown. There are therefore two types of chance. The first is that where we do not know the causes because they are hidden, and this is commonly called determinist chance. The second type of chance is indeterminist, one whose causes we do not know because they do not exist. The existence of this second type is denied by partisans of determinism or of the Theory of Hidden Variables. Double Causality considers that variables can be hidden by the fact that they are "futures", something which adds up to saying that chance does not exist.

Chance, dispersive: a particularly indeterminist hazard, because it is caused by the dispersion of its possibilities of evolution. For example, a rock tumbling down the side of a mountain is very dispersive because of all the rebounds it makes on different bumps on the ground (notably other rocks), that disperse its possible routes more and more unpredictably. With each rebound, its dispersion is amplified, and after a certain number of rebounds it becomes quite immoderate, making the hazard indeterminist, that is, resistant to infinite tiny variations in the initial conditions.

Chance, indeterminist: an event which in principle it is impossible to date the moment or to predict the position with any accuracy, even if we know exactly the initial conditions. For example, it is impossible to determine down to less than 100 metres or 3 seconds, the exact position that a car may have after driving 100 km down the motorway, even if it is fitted with a speed regulator and no obstacles slow it down.

Chaos Theory: the property (and the theory that describes it) of a dynamic system whose behaviour is so sensitive to the initial conditions that its state becomes completely unpredictable and indeterminist.

Coincidence: a chance meeting, or simultaneous events that are by nature improbable and bear some resemblance to each other.

Conditioning: a term used to describe a mental state of a being that gives the illusion of freedom while the evolution of the being is in fact determined in advance by often unconscious processes. The idea that we are always conditioned is amplified by the discovery made by neuroscientists that our conscience is itself a product of neuronal activity at work in our brains.

Cycle of love: the cycle of love is that of second causality. It consists of a retroactive loop between the present and the future, characterized by the need to wait for time to work or for the universe to work in our stead in order to realize our intentions automatically. The different stages of the cycle of love are desire, wish, intention, attention, faith, confidence, aspiration and finally joy. The first four stages correspond to self-sacrifice, which turns love into potential energy, investing the future while at the same time giving us a sense that it is disappearing. The following four stages correspond to the wait, while time or the universe do their work by making love come back to the present, having "flowed" from the future towards the present.

Determinism: a philosophical concept according to which the evolution of all physical phenomena is predictable using causality, and can in theory be calculated in advance if we know all the initial conditions.

Determinist: used to describe a system whose ulterior evolution is entirely fixed by well-established laws.

Dimensions: directions of space-time. Our physical space-time possesses four dimensions, towards the north, towards the east, up and towards the future, making three spatial dimensions and one temporal. According to String Theory as well as our own work, however, our true space-time must have other dimensions: six extra spatial dimensions, which instead of being straight directions, would rather be directions folded in on themselves, extremely tiny and as a result, invisible.

Entropy: a quantity measuring the disorder in a given whole of atoms or molecules. According to the second law of thermodynamics, the entropy of an isolated system can only increase. It follows that according to this law, the disorder of the universe can only increase with time, but this conclusion is the subject of fierce discussion among physicists.

Free will: a philosophical hypothesis according to which we are truly free, as human beings with conscience or mind, and thus capable of self-determination, in spite of our conditioning. Freedom is the basic assumption of the theory of double causality. It follows that our possible futures are as countless as branches on a tree, which is why we describe the freedom we have of controlling our future using the metaphor of the Tree of Life.

Gravitation (law of): the law of mutual attraction of solid objects whose intensity is inversely proportional to the square of the distance between them.

Graviton: a hypothetical particle that transmits the force of gravitation, actively sought by physicists using particle accelerators, but never yet observed.

Hidden variables (Theory of): the existence of hidden variables is invoked by the detractors of the majority interpretation of quantum mechanics whereby it is indeterminist: it is impossible to predict what state a particle will take when it is observed, this state being dependent on absolute chance. The partisans of hidden variables, such as Einstein, refuse to accept this indeterminism, and consider that it is only apparent, and will be explained by physics when we are able to explain "hidden causes" and more generally, the causes of all "apparently" indeterminist systems.

Indeterminism: as opposed to determinism, a philosophical concept that denies the fact that any event can be predictable using laws of physics. The principle consequence of indeterminism is the existence of hazards, that is, of multiple possibilities for the evolution of a system.

Indeterminist: is said of a system that is particularly subject to indeterminism, that is, presenting hazards that on principle render any prediction of its evolution impossible.

Initial conditions: the state of a dynamic system or hazard at the beginning of its evolution.

Inner dimensions: according to the theory of double causality, the six extra dimensions of space, invisible to our eyes, must be inner dimensions of our psyche (three for making the choice of the path and the three others for making the choice of the destination), whose purpose is to guarantee our free will, this latter expressing itself by moving along these extra dimensions.

Inner space: a concept of double causality that attributes to our psyche the extra, invisible dimensions of space time which make possible the parallel universes (interpreted as potential ones), and a capacity to displace us in these "exterior" parallel spaces. Passing from one space to another results from a choice made at the moment of observing an alternative between two possible choices, or a bifurcation.

Interaction: to explain the behaviour of matter and of different states, notably solid states, physicists call on four fundamental forces or interactions: strong, weak, electromagnetic and gravitational.

Inverted determinism: determinism in the reverse direction of time, based on the idea that the past is determined by the present thanks to traces of the past and the Law of Converging Parts. Inverted determinism would thus be more determinist than classical determinism, inasmuch as traces of the past are much more numerous and reliable than traces of the future.

Irreversibility: irreversibility is a very general characteristic of evolutionary phenomena observed at our macroscopic level, and translates into the obvious impossibility of evolving in the opposite direction. However, irreversibility is absent from the fundamental equations of physics, which means that consequently this is an irreversibility on fact, not on principle, that corresponds to reversibility in principle, but with an extremely low probability.

Law of Converging Parts: a hypothetical, missing law based on inverted determinism that allows us to calculate the past from the present, or the present from the future. This law is founded on the reversibility of the fundamental equations of physics and places itself as a law opposed to the law of increasing entropy: it produces order instead of producing disorder.

Law of universal attraction (of life trajectories): the general application of the Law of Converging Parts to "life trajectories". Taking into account the omnipresence of the future and the past, when a modification happens on a life trajectory, it is instantaneous all along the whole trajectory, including that part that only exists in a state of potential as yet unrealized. This is why the law is dynamic and results in bringing together (attracting) trajectories linked by convergent intentions. This "attraction" is supposed to be a perfect analogy with the law of universal gravitation which acts on the trajectories of material (solid) objects.

Law of universal gravitation: law established by Newton then generalized by Einstein, describing how the trajectories of solid objects attract each other, respectively indirectly proportional to the square of the distance between them and proportional to the curvature of space-time.

Love: the matter or energy of the mind, memorized in omnipresent space, outside of the present time, in the form of probabilities for actualizing possibilities. The greater the love, the higher the probability for actualizing the object of that love. However, we must confuse love with the different stages of its cycle, like desire, for example, which would be susceptible of lessening its effects by preventing it from circulating (see cycle of love).

Magic: describes a remarkable phenomenon that is impossible to understand in the context of causality, because it appears to violate the known laws of physics. Magic is a product of second causality and it is amplified by the circulation of love.

Materialism: a philosophical doctrine affirming that nothing exists outside matter and that the mind is itself entirely material.

Materialist: a partisan of materialism.

Metaphor: the modification of the meaning and representation of a concept in the form of an image or a comparison with the object of making it simpler to understand the original concept.

Metaphysics: philosophical research that aims for knowledge and understanding of the "being".

Mind: the part of the being or conscience that is not determined by the processes at work in our brains. The mind sits outside the present in all the omnipresent space of our possible futures (and/or in our past). It is displaced by our intentions and its ability to influence our present is determined by the capacity of the being to "love", love being supposed to be the essence or fuel itself of the mind, its fluid "matter", its energy, present in space in the form of probabilities.

Multidimensional: used to describe a space with several dimensions*, usually more than four.

Murphy's Law: an empirical principle that states that if something can go wrong, then it inexorably will go wrong, especially when we most need it not to.

Observer: the role of observation that consists of making a single version of the world or of an individual evolution enter reality, from all the multiple possible versions that exist simultaneously before being observed. This role thus implies the distinction between reality and existence, reality being by definition unique, whereas existence may only concern non-lived, that not yet observed, realities. The role of observer is a generalization of the role of the observer in quantum mechanics.

Omnipresence: the act of being present in different places but always at the same time.

Paradigm: a way of conceiving the world that is shared by the members of the same scientific community and sometimes by whole societies.

Parallel universes: universes that exist in parallel but "apparently" disconnected from our own, that is, inaccessible to observation by us. Quantum mechanics and String Theory both envisage the existence of such parallel universes.

Potential: used to describe a possible but not as yet realized life trajectory. A potential is measured by its probability for realization.

Quantum mechanics: a branch of physics that describes the behaviour of atoms or particles and their interaction with light, by making probabilities play an essential role. At any moment, atoms only exist in the form of superpositions of all their probable states. This superposition of possible states is also called wave function. Observation of the state of a particle using an instrument forces it to take on a single state, this phenomenon being called "wave function collapse (reduction to a single state)". The philosophical implications of quantum mechanics are considerable and have contributed to a progressive modification in physicists' vision of the world during the 20th century.

Relativity (Theory of): a geometrical theory developed by Einstein in 1905, establishing a close connexion between space and time. These notions depend on the motion of the observer and the notions of simultaneity and present lost their meaning: everything is relative to the position and speed of the observer.

Retrocausality: causality in the reverse sense of time, a concept initially developed to justify the apparent observation of particles travelling back in time. However, the concept is contested by physicists who have introduced antiparticles as an alternative justification, confirmed by experiments. Not to be confused with "second causality" which is also causality in the reverse sense of time, but this is determinist and does not suppose that its effects "precede" its causes.

Reversibility: the principle at work in all fundamental equations in physics, making them symmetrical in relation to time, which implies that all phenomena should be able to evolve backwards as well as forwards, even though this is not what we observe in nature. Other equations in physics, valid solely at macroscopic level for a large number of particles, do, however, explain this irreversibility. These empirical equations bring in chance and probabilities, and are a product of statistical physics.

Spiritism, spiritualist: a doctrine which holds that man is composed of a perishable body and an immortal spirit capable of removing itself from the body and surviving it.

Statistical physics: a branch of physics that aims to explain and predict the behaviour and movement of physical systems containing a large number of microscopic particles: these systems are themselves termed "macroscopic". Calculations in statistical physics are characterized by the fact that they use the massive intervention of chance in the choice of particle trajectories after collision.

String Theory: a theory based on the idea that elementary particles are not points but infinitesimal pieces of string (10^{-33} cm), thus giving extra, invisible dimensions to space.

Synchronicity: a significant coincidence, that is, one whose mysterious nature carries meaning. It is characterized by the fact that the psyche of the person involved seems to be all the more implicated the lower its probability, leading us to think of a "sign of destiny".

Talking with Angels: a phrase describing the inner dialogue between the being situated in the present and the other part of the being generally situated somewhere in the future, at the focal point for the intention. This dialogue exploits a retroactive loop of second causality that allows information from the future to be sent back to the present, in answer to requests made in the present, through the intermediary of intention deposited or planted in the form of questions asked of the Angel. This dialogue may be supported by all natural phenomena, in particular those subject to hazard. We also talk of a "dialogue with minds".

Tree(s) of Life: a metaphor for the theory of double causality, using a path along the branches of a tree to represent our different, omnipresent, future potentials.

ANNEXE

Indeterminism and TDC

To illustrate the indeterminism of nature, let us take the example of a rock, tumbling down a mountain slope. The rock's trajectory is rendered dispersive by the multiple bounces it makes over the uneven ground, over vegetation and other rocks. After each bounce, the extent of the possible trajectories of the rock is further amplified, making its path more and more unpredictable. After a certain number of bounces, it becomes impossible to make the least prediction about the rock's position, unless one knows with quasi-infinitesimal precision both its initial position and the position of every bump in the ground.

Now, the amount of precision we can have about the initial conditions of any material object, in time and position, is limited by the Heisenberg Uncertainty Principle in quantum mechanics. In nature, every happens as if space were quantified and its geometry discontinuous, as if as a result, to describe microscopic phenomena smaller than a certain scale, we were forced to call on probabilities. At such tiny scales, it is fundamentally impossible simultaneously to know the position and speed of a particle. All knowledge of initial conditions is thus on principle limited below a certain distance or a certain length of time.

The already well-known philosophical consequence of quantum mechanics is that at microscopic level, fundamental indeterminism reigns. It is in fact so impossible to predict the future of a group of states of matter at this scale that physicists have concluded that all possible states permitted by the laws of physics must exist simultaneously. Therefore there is no single future, but an infinity of future potentials. On the other hand, the fact of observing a state by measuring it with an instrument immediately results in its being determined and rendered unique. So it is only thanks to our observations that the reality we know only comes in a single version. Without observation by the human beings we are, physics would describe a world where an infinity of realities coexisted simultaneously in the form of parallel universes. So it would seem that we only know "one universe" because we are gifted with a conscience for observing it. This is what makes many "spiritualist" physicists say that we are co-creators of the world around us.

The Theory of Double Causality (TDC) does not go quite so far, being closer to the opinion of "materialist" physicists who consider that our conscience has no effect on the creation of reality.

To explain the creation of a unique version of reality, TDC first appeals to the idea that indeterminism reigns everywhere in nature, not just at microscopic, but also at macroscopic level. This assertion is based on the generalized existence of indeterminist accidents in the world around us, our falling rock being one example.

In order fully to understand this assertion, let us return to our rock. Given our observation on the phenomenon of the increased dispersion of its trajectory, it is quite reasonable to imagine that multiple scenarios could exist for the rock's fall, and after falling down 100 metres, it might end up in any one of several places some 10 metres apart, for example, even though its initial departure point had been the same. The simple indeterminist details controlling the rock's position at a microscopic level, at the moment it starts rolling down the mountain, are supposed by TDC to increase to such a point that they engender an important amount of macroscopic indeterminism about the position it will reach 100 metres further down.

To find a mathematical justification for this observation, let us replace our falling rock and mountain with a billiard ball making several rebounds in a very special sort of billiards game, called "Sinaï Billiards". In this game, a fixed, spherical obstacle is placed in the middle of an ordinary billiards table, similar to a large ball of equal diameter to half of the width of the table. Here it will be replacing the bumpy ground our rock falls down. What is so interesting about such an example is that it allows us to make calculations, so we can demonstrate that this table is chaotic. As a result of this chaos, it is impossible to predict the position of a ball once struck, for any more than a few seconds. This phenomenon is due to the fact that after each rebound off the central obstacle, the uncertainty about the trajectory of the ball that has been struck is increased by the rebound. We can thus demonstrate that to predict the position of the ball to within ten centimetres after ten seconds, we would have to strike the ball with the precision to within two hundredths of a millimetre. We can also predict that the difference between two balls with an infinitely small difference in trajectory at the outset is multiplied on average by three every second. Now, if for example we start with the particularly large number of three to the power of sixty, we can deduce that in order to strike a second ball along a trajectory, that for a whole minute will diverge by no more than one millimetre to that of the first ball, then the level of precision necessary will need to be much, much smaller than the size of an elementary particle!

Ignoring quantum indeterminism, many physicists keep on labelling chaotic phenomena as determinist yet "unpredictable", as if the fact of having a trajectory that was unpredictable because of its chaotic nature did not prevent this trajectory from being determinist, given that in theory, it is possible to calculate it. This reasoning forgets two things, though, one being the fact that the precision of any calculation has to be limited, and the other being the fact that below a certain level of precision, matter already behaves in a fundamental indeterminist way. The hypothesis at the origin of this reasoning, which states that we live in a continuous space, is thus false.

According to TDC, qualifying a chaotic system as "determinist" only perpetuates an illusion produced by the idea that our equations are themselves determinist. Such an idea is based on the false hypothesis that we evolve in a continuous world, in which these equations could in fact only lead to a single solution. Quantum mechanics shows us that, on the contrary, we live in a fundamentally discontinuous world at a microscopic level, where no quantum object can hold a precise position. As a result, an unpredictable determinism becomes equivalent to real indeterminism.

When, according to other points of view, we accept the indeterminism of nature, because we have no means of predicting the future of a system or of understanding its causes, we often talk of chance. But according to TDC, chance doesn't exist in the real world, because when a phenomenon is not caused by the past, it is caused by the future. All that exists is our ignorance of the real causes or the hidden variables, whether they be past or future. When these causes are not due to the past and as a result we invoke chance, we are faced with a phenomenon that is by definition indeterminist, and for which physics provides us with no answer.

So it is indeed to guarantee the absolute coherence of the scientific approach, by eliminating chance and indeterminist accident, something vital for physics, that TDC suggests using future causes to determine whatever is not determined by past causes. But in order for this double causality to be well-founded, the evolution of systems in the inverse sense of time has to be determined by a missing law, called the "Law of Converging Parts", a law that is fundamental because it creates order.

According to a principle of maximizing created order, such creation of order would have the advantage of handing us a unique determinist solution for the past, even in the presence of insufficient traces. In effect, it is conceivable that, contrary to a law that creates disorder and forces us to exploit chance and statistics to avoid stumbling on multiple futures, a law that creates order could make just one solution emerge, the most ordered one, from all the solutions that respect the reversible laws of physics.

Thus all past indeterminism would be solved, not by an accumulated effect of our intentions as is the case with the future, but by a generalized law of universal attraction that would introduce harmony everywhere, resulting in each part showing a reflection of the whole.

BIBLIOGRAPHY

1. Stephen Hawking, Leonard Mlodinov, *Y a-t-il un grand architecte dans l'univers ?*, Odile Jacob, 2011.

2. Rémy Chauvin, *Le Retour des magiciens*, JMG Éditions, 2002.

3. Deepak Chopra, *Le Livre des coïncidences*, Interéditions, 2004.

4. Francis Crick, *L'Hypothèse stupéfiante : à la recherche scientifique de l'âme*, Éditions Plon, 1995.

5. Régis Dutheil, Brigitte Dutheil, *L'Homme superlumineux*, Éditions Sand, 1990.

6. Michel Granger et Jean Moisset, *Coïncidences, hasard ou destin ?*, Éd. Trajectoire, 2003.

7. Brian Greene, *La Magie du cosmos : l'espace, le temps, la réalité: tout est à repenser*, Éd. Robert Laffont, 2004.

8. Eugen Herrigel, *Le Zen dans l'art chevaleresque du tir à l'arc*, Éd. Dervy, 1961.

9. Pierre Jovanovic, *Enquête sur l'existence des anges gardiens*, Le Jardin des livres, 2005.

10. Pierre Jovanovic, *L'Explorateur de l'Au-delà*, Le Jardin des Livres, 2005.

11. Allan Kardec, *Le Livre des Esprits*, 1857.

12. Étienne Klein, *Le Temps existe-t-il ?*, Éd. Le pommier, 2002.

13. Étienne Klein, *Les Tactiques de Chronos*, Poche, 2004.

14. Étienne Klein, *Le Facteur temps ne sonne jamais deux fois*, Flammarion, 2007.

15. Arthur Koestler, *Les Racines du hasard*, Calmann-Lévy, 1972.

16. Ervin Laszlo, *Science et Champ Akashique*, Éditions Ariane, 2005.

17. Irène Magnaudeix, *Pierres assises, pierres mouvantes*, Les Alpes de Lumière, 2004.

18. Gitta Mallasz, *Dialogues avec l'Ange*, Aubier, 1998.

19. Michael Newton, *Souvenirs de l'Au-Delà*, Le Jardin des Livres, 2008.

20. Osho, *Être en pleine conscience*, Éditions Jouvence, 2005.

21. David Peat, *Synchronicité, le pont entre l'esprit et la matière*, Éditions Le Mail, 1988.

22. Emmanuel Ransford, *La Nouvelle Physique de l'esprit*, Éditions Le Temps Présent, 2007.

23. Roger Penrose, *L'Esprit, l'ordinateur et les lois de la physique*, Interéditions, 1993.

24. Ilya Prigogyne, *Les Lois du chaos*, Flammarion, 1999.

25. Dean Radin, *La Conscience invisible*, Presses du Châtelet, 2000.

26. Hubert Reeves et al, *La Synchronicité, l'âme et la science*, Albin Michel, 2001.

27. Hubert Reeves, *L'Heure de s'enivrer*, Éditions du Seuil, 1986.

28. Hubert Reeves, *La Première Seconde*, Éditions du Seuil, 1995.

29. Vlady Stévanovitch, *La Voie du Taï Ji Quan*, Dangles éditions, 2008.

30. Lynne Mc Taggart, *La Science de l'intention*, Ariane, 2008.

31. Jean-François Vézina, *Les Hasards nécessaires*, Éd. de l'Homme, Québec 2001.

32. Neale Donald Walsch, *Conversations avec Dieu 1−2−3*, Aventure secrète, 2007.

33. Trinh Xuan Thuan, *Le Chaos et l'harmonie*, Fayard, 1998.

34. Vahé Zartarian, *L'Esprit dans la matière*, Éditions Georg, 1998.

35. Vahé Zartarian, Martine Castello, *Nos Pensées créent le monde*, JMG Éditions, 2003.

36. Vahé Zartarian, *Le Jeu de la création*, Éditions 3 Monts, 1999.

37. Dr Hans-Joachim Zillmer, *L'Erreur de Darwin*, Le Jardin des Livres, 2008.

38. Michel d'Aoste, *Les Secrets de l'astrologie universelle*, The Book Édition, 2008.

39. Antoine Suarez, Peter Adams, *Is Science Compatible with Free Will?*, Springer, 2012.

40. Nicolas Gisin, *L'Impensable Hasard*, Odile Jacob, 2012.

41. Ilya Prigogine, *La Fin des certitudes*, Odile Jacob, 200 l.

42. Friedrich Nietzsche, *Humain, trop humain*, Pluriel, 2001.

43. Henri Bergson, *Le Possible et le réel*, PUF, 2011.

44. Thibault Damour, *Et Si Einstein m'était conté*, Cherche Midi, 2005.

45. Carlo Rovelli, *Et si le Temps n'existait pas*, Dunod, 2012.

46. Jean Staune, *Notre existence a-t-elle un sens?*, Presses de la Renaissance, 2007.

47. Jean Staune, *La Science en otage*, Presses de la Renaissance, 2010.

48. Philippe Guillemant, *A Flexible Cylinder to Model Physical Functions of Consciousness, Cosmopolis* N° 2, 2013.

49. Philippe Guillemant et al, *Characterizing the Transition from Classical to Quantum as an Irreversible Loss of Physical Information*", arXiv:1311.5349 [quant-ph], 2013.

50. Jacques Vallée, *Science interdite* Vol 2,Aldane, 2013.

51. Jacques Vallée, *A Theory of Everything (else)*, Conférence TEDx Bruxelles, 2011.

52. Olivier Costa de Beauregard, *Le Temps déployé, passé, futur, ailleurs*, Le Rocher, 1988.

53. François Martin, *Synchronicity*, film DVD de Jan Diederen, 2012.

54. W. Pauli & C.G. Jung, *Correspondance 1932-1958*, Albin Michel, 2000.

55. Joachim Soulières, *Les Coïncidences*, Dervy, 2012.

56. Eckhart Tolle, *Le Pouvoir du moment présent*, J'ai lu, 2010.

57. Alain Connes, Danye Chéreau, Jacques Dixrnier, *Le Théâtre quantique*, Odile Jacob, 2013.

58. Serge Haroche, *Jongler avec la lumière*, De Vive Voix, 2010.

59. Philippe Guillemant et al, *A Discrete Classical Space Time Could Require 6 extra Dimensions, Annals of Physics 388* (2018), 428-442.

CONTENTS